# Shiftwork Safety and Performance

## A Manual for Managers and Trainers

# Shiftwork Safety and Performance

## A Manual for Managers and Trainers

Peggy Westfall-Lake
Glenn N. McBride

LEWIS PUBLISHERS
Boca Raton Boston London New York Washington, D.C.

**Library of Congress Cataloging-in-Publication Data**

Westfall-Lake, Peggy.
Shiftwork safety and performance : a manual for managers and trainers / Peggy Westfall-Lake.
p. cm.
Includes bibliographical references and index.
ISBN 0-56670-257-7 (alk. paper)
1. Industrial safety--Handbooks, manuals, etc. 2. Shift systems--Handbooks, manuals, etc. 3. Hours of labor--Handbooks, manuals, etc. I. Title.
T55.W49 1997
658.3'82—dc21
for Library of Congress 97-29082
CIP



Direct all inquiries to CRC Press LLC, 2000 Corporate Blvd., N.W., Boca Raton, Florida 33431.


International Standard Book Number 1-56670-257-7
Library of Congress Card Number 97-29082
Printed in the United States of America 1 2 3 4 5 6 7 8 9 0
Printed on acid-free paper

# Authors

**Peggy Westfall-Lake** is a Human Factors Consultant, with Williams Learning Network in Houston, Texas. She is an author, safety researcher, training designer and distance learning curriculum developer. Ms. Westfall has had experience consulting with a variety of 24-hour industries, union and non-union, including: manufacturing, pharmaceutical, mining, nuclear fuels processing, emergency services, law enforcement and others. Her research has covered the impact of downsizing on safety and human performance, promotion of alertness and vigilance, improvement measurement criteria selection, schedule change processes and prevention of human error. A Toastmaster presenter, Ms. Westfall is a popular keynote speaker at regional, state, and national level safety and health conferences.

**Glenn McBride** is President of McBride Publications. As a training consultant, author, and corporate training developer, he is best known for creating breakthroughs in behavioral safety, 24-hour communications and team safety cultures. Mr. McBride (B.S., M.A. in Behavioral Sciences) has twenty years of experience in chemical, hospital, and military shiftwork and built a career with the Dow Chemical Company as a plant technician, supervisor, manager, and training designer.

# Acknowledgments

Getting this manual and Performance Tracking and Improvement into a usable form took incredible encouragement and support over a three-year period. The project was catalyzed by Ted Swenson of The Dow Chemical Company who gave us "one more corporate project" before retiring to a Colorado mountain hideaway.

Jon Walker and Robert Rogers were supportive Dow Chemical training managers who patiently kept asking when the original Dow shiftwork program would be finished. Gene Lowe was the Dow representative and former plant shiftworker who surveyed hundreds of corporate shiftworkers around the country, then supported, nurtured, and piloted the original two-hour TIME TRAVELERS, F.A.S.T. Tracking program.

We owe a tremendous amount of human insight to Charlie Kilsby, a local industrial Gulf Coast barber, who has interviewed more 24-hour employees — albeit male — than nearly anyone else we knew. Mr. Kilsby has added a personal depth to our research that we could not have achieved otherwise. Also:

Bob Nute, NASA Flight Activity Operations; Diana Ormsbee, NASA News Department; and Vincent J. Mellone, ASRS Deputy Program Manager, were instrumental with their advice and technical assistance.

The customers, who risked running the F.A.S.T. Tracking program early in the game which gave us tremendous encouragement and led to the writing of this manual.

Drs. Graeber, Merten of the FAA, and Hauri of the Mayo Clinic, and the many other researchers who gave us much needed materials, research data, and encouragement.

# Preface

This book is designed for Safety Directors, Human Resource Managers, Training Managers, Shift Supervisors, Production Managers, and Plant Managers of 24-hour employees - shiftworkers. These are the working people who commit to "time travel" around the 24-hour clock. The message to management within this book should be considered a "safety wake-up call."

Although the information and recommendations contained in this publication have been compiled from reliable sources, we can make no guarantee as to the current status of certain data. Other or additional safety measures may be required under particular circumstances to promote concepts contained herein. But, we can assure you that promotion of employee health, fitness, alertness, sleep improvements, and quality of time off will have a positive effect on your company's safety and productivity.

This book is really a training manual plus program generator! We would like this book to speak through you — to the important people in our board rooms, control rooms, plants, cockpits, and departments, who will through their management, skills, ability, and 24-hour performance, lead us into the twenty-first century.

Peggy Westfall
Glenn N. McBride

# How to Use this Book

This book contains key concepts and data that you will want to share with both management and 24-hour (shiftwork) employees. Here are some sessions you can quickly plan and deliver from the contents of this book:

**Watch for the overhead projector icons.** Every time you see the icon by an illustration, you can use the enclosed disk to assist with your own presentations.

To present the importance and seriousness of shiftwork training to your management ... talk about the critical 24-hour jobs, Set-Ups and Slips, the two shiftworker models, and safety problems. Use content from Chapters 1, 2 and 3.

To energize and motivate your employees ... review the critical job and extra-achiever ideas with them. Tell the Edison story. Review the content from Chapters 1 and 2.

To begin a Performance Tracking and Improvement Program ... review the F.A.S.T. Tracking techniques, then cover how to fit them into targets, efforts, and results. Challenge your employees to start by trying something small. Review Chapters 2, 5 and 6.

To brainstorm new rotating schedules, select schedules and concepts from Chapter 3.

Shiftwork safety and performance are moving, progressing targets! You must re-assess and measure your schedule change factors and performance improvements periodically. Changing social and economic factors can quickly negate your best efforts, along with restructuring and downsizing.

# Table of Contents

# CHAPTER ONE

## Performance Improvement and Critical Jobs

### OBJECTIVES OF CHAPTER ONE

You will be able to describe the relationship between Technical Empowerment and critical 24-hour jobs.

You will understand that changes to shift schedules or implementing a performance improvement program should be done without assessing all areas to be impacted.

After you finish the first chapter of this book, you will see that 24-hour operations carry a unique load on the human performer. Technical empowerment and pressure on employees will only increase, along with set-ups for potential human error.

### CAN YOU IMAGINE ...

To feel the importance of our message, we usually begin our management training session with the following...

Imagine that you've won the lottery ... a big lottery, of course. And with your fortune, you've decided to purchase a winning professional football team. In fact, since you own the team, you've selected yourself to be the new head coach!

"Oh, what a manager's job!" you're saying as the full impact of your responsibilities hits you. Just a minute though, the job turns out a little differently than you anticipated.

As you enter the main training facilities at 9 a.m. on that first Monday morning, you stroll by a large room to the right with twenty staff members doing aerobics. Further down the hallway off to the left, you overhear a group of office professionals talking about the new high alertness and weight control menus in the cafeteria. As you near your brand new office, you pass a conference room with a committee talking to a hard-bodied fitness consultant, apparently planning the next quarter's physical activities for the managers and office staff.

You can hardly wait! With all of this office pro-active health activity, your new football team must have a heck of a fitness and performance improvement program. Eager to meet the team, you drop into the office of your number one offense coach and introduce yourself and say, "Wow, Charlie! I can't wait to meet the team and see the team's training schedule."

So Charlie takes you down the hall, down to the big double doors that say, *Coaches and Team Only*. He opens the doors and there are your team players sitting on benches with fast food bags crumpled in all shapes, sausage and biscuit sandwiches, fried potato breakfast wedgies, soft drinks, croissant sausage sandwiches and fried apple pies. High-dollar players looking a little over-filled and over-saturated.

You stand there shocked ... "Charlie," you say, "what's the deal? Where's the performance diet — the wellness program — the aerobics for the players?"

Charlie responds, "Well, we've thought a little about it, but right now only the managers and office staff get the aerobics and fitness training. Some of the players who are performing poorly get *some* performance improvement training. It's just not in the budget for everyone. Some of the better team members workout and watch their own fitness a little, but it's strictly something they do on their own. Everyone is expected to learn the procedures and plays. It's what we call our new League Compliance Program."

Just imagine! How would you feel? How do you think your team will perform this next season? Think you've got a coach's job cut out for you? You bet! Are you ready to wake up from your dream ... or do you want to stay and re-educate your coaching staff and try to save the team?

Okay, so you probably don't own your own football team and you're probably not the head coach ... but you're not dreaming the unusual. Unless the team you work with is one in a million, it is in the exact same situation of those imaginary football players. This is what we call the "minimum compliance syndrome" in our 24-hour workers. The term "performance improvement program" is often reserved for those problem cases where employees perform unsatisfactorily!

Training attention tends to become focused on complying with Federal Regulation; there are less resources to apply to front-line performance improvement. Even management and office staff seem to find the resources to address personal performance improvement.

**The Minimum Compliance Syndrome in 24-Hour Operations**

| | |
|---|---|
| ■ Teaches Procedures and Plays? | YES |
| ■ Teaches Precautions and Hazards? | YES |
| ■ Satisfies Min. Government Compliance? | YES |
| ■ Considers Ergonomics Issues? | YES |
| ■ Counts Safety Incidents? | YES |
| ■ Provides Fitness Facilities? | YES |
| ■ Teaches Vigilance Techniques? | NO |
| ■ Promotes Fitness For Operational Duty? | NO |
| ■ Promotes Personal Performance Improvement? | NO |
| ■ Optimizes Scheduling for Human Factors? | NO |

In operations with fitness facilities, we occasionally find the doors locked at 6 p.m. until 6 a.m. when the office staff arrives. What kind of signal does this give the night shiftworker? High fat snacks and "foods" are kept stocked in the vending machines for workers needing nourishment at 3 a.m. How alert will they be at shift change after their 90g fat burritos and 35g fat pastry?

The plant nurse comes on shift at 8:30 a.m. to begin "wellness program" blood pressure and cholesterol testing. Your night shift gets off work at 7:00 a.m. Do you think they'll wait around to begin their program? Where is their performance improvement program? How do you get your front-line operations motivated and actively involved with changes to promote better alertness, fitness, health and energy? This book will help answer those questions and give you some ideas about where to find opportunities for front-line performance improvement.

## PERFORMANCE IMPROVEMENT IN CRITICAL JOBS!

Fortunately, training and performance improvement are very high priorities in some jobs in this country. What about fighter pilots? Reconnaissance troops! Shuttle astronauts! What if there was no alertness, fitness or general performance improvement push in these critical jobs?

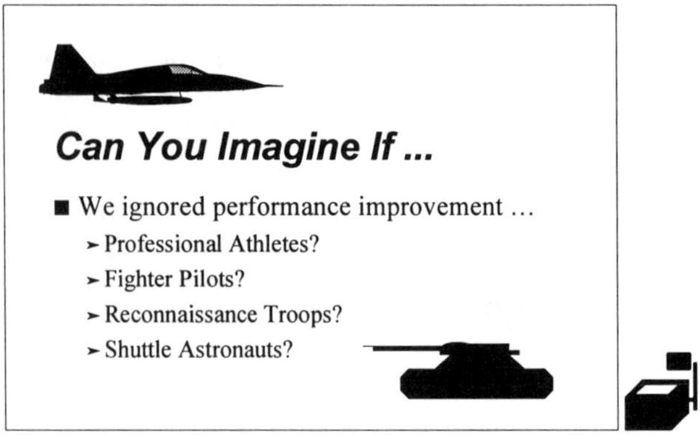


And yet, this very omission is occurring across the United States. Everywhere we see American employees working critical 24-hour jobs without continuous performance improvement ... those jobs where it should be absolutely necessary to be in top notch readiness and condition. Those jobs where acuity of judgment must be 100% ... 100% of the time! A much higher percentage of shiftworkers hold critical jobs than do dayworkers. This doesn't mean our supervisors and managers don't have important jobs; what it means is that the opportunity for cataclysmic disaster in many cases is in the hands of the people who work on-shift — people who can suffer jetlag and extreme fatigue. For these critical jobs, performance tracking and improvement at the individual level should become a job style.

**_The Majority of 24-Hour Jobs are CRITICAL!_**

A much higher percentage of shiftworkers hold critical "technically empowered" jobs than do dayworkers!

Country-wise, this issue has become very important.[1] In the 1990s, about one in five American workers is working a 24-hour job.[2] A shiftwork job. This number is increasing according to the Bureau of Labor Statistics. If you look closely, 90 percent of these jobs are critical ... chemical plant and refinery workers, nuclear plant operators, police officers, mill and food processing workers, truckers, pharmaceutical workers, airliner crews, train crews,[3] emergency and medical crews, mining operations, and numerous other job positions. These jobs are critical to the safety and security of our plants and factories, our communities, our lives (as emergency rescue teams), the stability of our society, and safety of our environment.

## Technical Empowerment and Critical Jobs

Technical Empowerment is our term for the new high potential jobs around the country that are propelled by technology. These critical jobs in the nineties have the potential for accomplishing great good through technology ... and the potential for creating tremendous damage and destruction. Individuals working these jobs have very important jobs.

**Majority of 24-Hour Jobs Are Critical**

- Chemical Plant Workers
- Refining/Oil/Gas Movement
- Nuclear Plant Ops.
- Mill/Food/Pharmaceutical Ops.
- Police Officers
- Airline Crew Members
- EMS/Hospital Staffs/Firefighters
- Maritime

Shiftworkers in critical jobs are now literally holding our businesses, communities, and human bodies together 24 hours a day. Just think about their impact! However, as managers and trainers, we spend copious amounts of time promoting their technical skills, yet we spend very little time promoting their mental, physical, and lifestyle performance.

This book is all about promoting the absolute best human performance in these 24-hour people ... and reducing the human error rate during that job performance. Optimizing performance and eliminating error go hand-in-hand!

Perhaps one of the most important purposes for writing this book is to emphasize that changes to shift schedules or improvements to the shiftwork lifestyle should not be done for any ONE of the following reasons:

**Promoting Shiftwork Improvement ...**

- Not for Any One REASON!
  - Increase productivity
  - Improve alertness
  - Improve teamwork and morale
  - Reduce accidents
  - Lessen corporate insurance costs
  - Promote employee lifestyle
  - Improve bottom line

Efforts to make changes to promote any one of these reasons will backfire. Rather, as a manager or trainer of shiftworkers, we strongly feel that it is your job to help your shiftworkers optimize their performance and reduce error FOR ALL OF THESE REASONS! Meantime, it is our job to find the best ideas and advisors around the country and make them available to you.[4]

## ELIMINATING SHIFTWORK SET-UPS AND SLIPS!

In this book, we will give you methods of promoting 24-hour safety and performance with your people. We'll also help you and your shiftworkers identify "Shiftwork Set-Ups." Shiftwork set-ups occur daily around the world. These "set-ups" lead to "human error slips" which cause collisions, spills, gas releases, fires, explosions, injuries and fatalities on a daily basis.

Set-ups that cause Slips can cost your company human pain and tragedy, multi-million dollar lawsuits, get you a bad headline in the newspaper, or a spot on the local news. (Someone once said in a training session that we should issue managers "TV news spot prevention tools!")

The human error occurring due to shiftwork follows a simple pattern: Set-Up to Slip to Incident.[5]

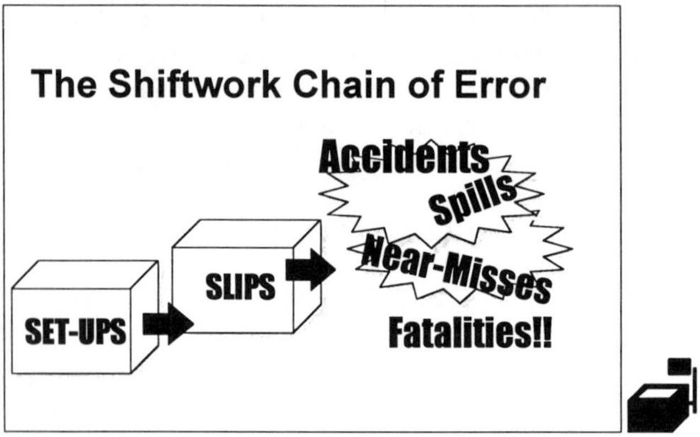


**The Shiftwork Chain of Error**

Let's work backwards from the incident and focus on the Shiftwork Slip.

*The Slip*

The slip that causes the incident may very likely occur late at night (as we will explore) but may happen at any time of the 24-hour day. The Set-up may have begun affecting the chain of events hours earlier.[6] The two types of slips we will discuss are physical and mental. Physical slips appear as fumbling, mis-labeling, mis-sorting, or mis-operating.

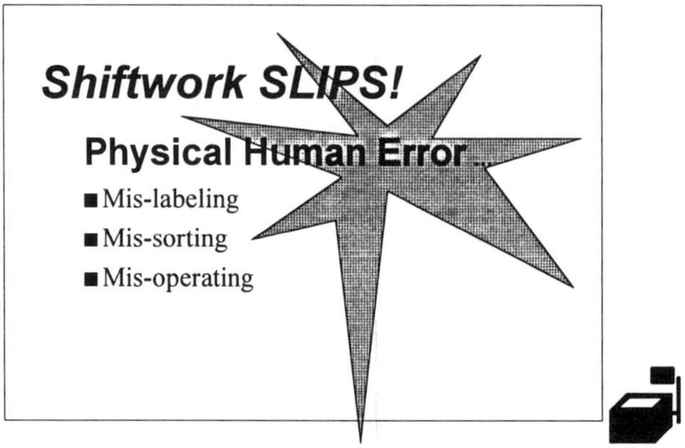


Mental slips or attention errors appear as misjudging, switching attention, accidental micro-sleeping and macro-sleeping,[7] or forgetting.

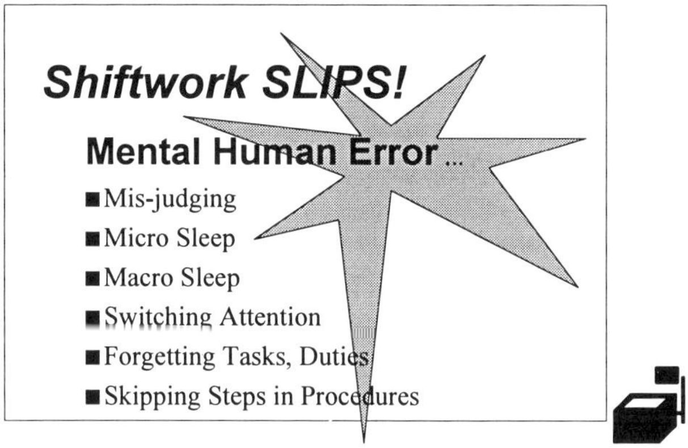


The slip is often pinpointed during the incident investigation. However, it is usually preceded (caused) by one or more set-ups that may be overlooked or forever debated afterwards. The shiftwork set-up is more difficult to implicate on individual incidents, unless great time and money is spent on the investigation and experts are consulted.

*The Set-Up*

Shiftwork set-ups that lead to slips include poor fitness and health, low alertness meals,[8] insufficient exercise/low body temperature, and excessive use of stimulants. All of these affect energy level, alertness, problem solving, and reaction time around the clock — and especially during the wee hours.

Other shiftwork set-ups are in the area of poor work/shift, sleep, and recuperation scheduling. These also affect energy and alertness.

Another high-tech set-up (often falling under a separate "engineering set-up" category) is the monotonous work environment where computers, radar, vehicles, and machines drone continuously through the shift.

Poor off-time activities and relations are also set-ups that impair performance, decision-making, and reaction time.

**Setups for Shiftwork-Related Human Error**

- Poor Work Schedules — Excessive OT
- Inappropriate Meals
- Insufficient Exercise
- Stimulant Abuse
- Poor Sleeping Environments
- Poor Sleep Schedules
- Monotony
- Distressful Family Life / Off-Time

In the section, Shiftwork and Safety, we give instances of shiftwork set-ups that have generated devastating consequences.

## GLOBEZILLA — AN OVERVIEW OF THE SHIFTWORK SET-UP

We now come to a bird's eye overview of the shiftwork set-up. Many contemporary factors have played special roles i.e., single family households, increased financial demands, downsizing, increased production demands, new team roles, global competition. Constantly impinging on the shiftwork set-up and 24-hour workers is the constant presence of Technical Empowerment — jobs that are ever-increasing in catastrophic potential.

**Globezilla**

**Set-Ups in the Middle of Cutting Edge Technology!**

You would think that in a nation of such critical jobs, cutting edge technologies, computer monitoring, and multi-million dollar hardware and software investments, these shiftwork set-ups would not be tolerated. **Unfortunately, it's rare to not find them in our plants, planes, hospitals, and police departments.** More than likely, your work team is running into several of these set-ups on every shift. So there is plenty of work for you to begin ... and soon!

Eliminating shiftwork set-ups is the place to begin a performance improvement program. This leads to fewer slips, improved morale, and better general performance.

**Set-Ups Due to Global Stress!**

Coming from the stress angle, we can explain how critical jobs, cutting edge technologies, and set-ups come together. When we began surveying managers and trainers, their candid remarks about shiftworker stress[9] due to "GlobeZilla," [10] rapid change, and downsizing "globally" overwhelmed us.

The following is the original "24-Hour Checklist" we published. Review it and imagine the response we received!

## Manager's 24-Hour Checklist

The following checklist was compiled while interviewing hundreds of supervisors in the manufacturing industry.[11] Look at these statements and see if you can identify with them. Copy them and try them on your managers!

### Global Stress and Competition:

[ ] Your company has recently "early retired" or laid off personnel. You wonder if the organization can make it safely with the current number of employees.

[ ] During business upshifts, your shiftworkers work extra overtime to cover for missing personnel. You sometimes wonder about their fatigue level.

### Downsizing and Job Pressure:

[ ] Your shiftworkers are under pressure to "get it finished now." When you are at home and they are at work, it sometimes occurs to you that they may be cutting corners to get the work done on time.

[ ] Your employees are working for a new shift supervisor due to recent organizational changes. You wonder if everything is under control.

### Personal Pressure:

[ ] Your shiftworkers are living in an age of family upheavals, divorces and other family problems. You wonder if this personal pressure is not affecting their attention to hazards of the job.

[ ] Your employees are slipping further under the gun of economics. More and more, you feel that they may tolerate anything to maintain pay level and pay off their debts. You wonder if this is a safe situation, and whether they will speak up if they are too fatigued to perform the job well.

### Technical Empowerment:

[ ] Your employees are working with larger and more destructive "guns." The machinery (towers, boilers, furnaces, reactors, ships, trucks, planes) are more automated, yet have more destructive potential than ever before. You see shiftworkers (and dayworkers) within your organization that have the "technical empowerment" to cause a major disaster. Certain night-related incidents stand

out in your mind, incidents like Chernobyl, Bhopal, Challenger, Three-Mile Island.[12]

[ ] In addition to the technical empowerment, your employees are being given more and more autonomy and allowed to make more team decisions.

[ ] Also, shiftworkers in your organization are working with less on-site supervision. You are concerned about this.

## A NEW APPROACH

**The problem is serious, however solvable!**

We, along with managers and safety experts around the globe, are beginning to grow to a new level of understanding ... an understanding that **Shiftwork set-ups and the resultant slips can be eliminated — certainly greatly reduced**. Amazingly, the actual elimination process that we will encourage you to implement isn't a gloom and doom, rules, regulations, and cautionary process. It's a proactive combination of preventative actions and countermeasures.

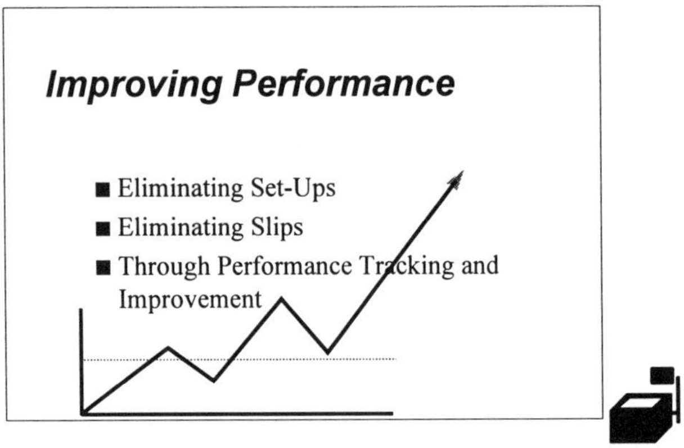


For a new solution we turn to a new approach! In an exciting new mode, we will give you valuable ideas on how to implement a **Performance Tracking and Improvement approach** across your entire organization. Performance Tracking and Improvement works for NASA, for our best athletic players, and for behavioral scientists facing tough behavior modification tasks. It will work brilliantly for your 24-hour performers also.

Whether or not you implement the full F.A.S.T. Tracking Performance and Improvement System, you'll be able to develop elements of this safety and performance program for your 24-hour people by the time you finish this book.

## SUMMARY — OUR GOALS FOR YOU

Therefore, our goals are quite straightforward. After reviewing this manual, you should be able to promote a Performance Tracking and Improvement Program for your shiftworkers — individuals in critical jobs working shifts other than straight 8 a.m.-4 p.m. days. As this program takes hold with your employees, they, through employee involvement and through gaining more individual control — will eliminate many of the shiftwork set-ups that we have already listed, discover set-ups unique to their operation, and begin to improve their own performance on and off-shift.

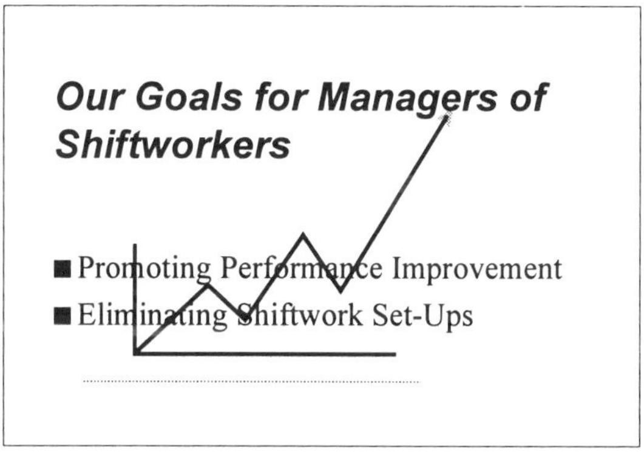

Soon, you will be able to measure improvements. You will see how you can help promote these improvements by making targets, attacking the set-ups and measuring success.

- gains in productivity
- improved alertness
- improve teamwork
- reduced accident rates
- reduced corporate insurance costs
- better employee lifestyle
- improved bottom line.

Before we get into performance improvement and current thinking about scheduling, we must explore why some shiftworkers do extremely well and why others barely seem to make it! So let's take a look at two views of the shiftworker.

**CHAPTER ONE ENDNOTES**

[1]"Depending upon one's source of information, there are now 14 to 25 million people in the United States whose work calls for them to rotate between day and night shifts." From: *Biological Rhythms and Rotating Shiftwork — Some Considerations for Air Traffic Controllers and Managers.* DOT/FAA/AM-86-2, Final Report by Carlton E. Melton, Ph.D., Office of Aviation Medicine and Robert S. Bartanowicz, Ed. D., Air Traffic Plans and Requirements Service, January 1986.

[2]From: Peter J. Sheridan, Executive Director of Occupational Hazards Magazine in *Occupational Hazards*, November, 1991.

[3]See *Transportation StarTrek in the 1990s* in the Shiftwork and Safety section.

[4]We fully support 24-hour careers. Although some dayworkers would say that shiftworkers work "second-rate schedules," we do not agree with that by any means! The best things simply take more than eight hours once a day: jet planes, multi-million dollar processes, police enforcement, medical, rail craft, the nation's trucks, education, research, and even writing training manuals. An important goal of this manual is to convince managers that the average shiftworker is an extra-achiever — a person who begins his/her career intending to achieve more than the average dayworker.

[5]See *Incidents and Costs* in the Appendix for reported examples of incidents and estimated costs. You may know the details of several of these incidents, many of which were due to human error Set-Ups and Slips.

[6]A classic case of long-term Set-ups leading to judgment slips is the Challenger spacecraft incident where time pressure, overtime and fatigue were involved.

[7]Micro-sleep is short, under three-minute unintentional "noddings." (Some experts define micro-sleep as those periods of mental "drop-outs" that last up to a few seconds with the eyes remaining open.) Macro-sleep is sleep lasting more than three minutes.

[8]Low alertness meals include: high fat foods, sugars, and starches — and simply whenever over-eating occurs. We do not believe that sweets increase one's alertness as you will see in the "alertness eating procedure."

[9]Northwestern National Life Insurance Company's Peggy Lawless surveyed 1,299 general employees from randomly selected companies. The study found that in 1992 thirty-nine percent were thinking about quitting their jobs because of the pressures. Four out of 10 said their jobs were very or extremely stressful. Thirty-two percent felt they would burnout in the next year. From: Employers ignore stress in workplace, study says by Cynthia Shanley, *Houston Post*, May 17, 1992.

[10]GlobeZilla, the monster of global stress and competition, has made quite a point about the origin of employee stress in our Dow Chemical safety programs since 1989.

[11]From: Glenn McBride's safety discussions and surveys with supervisors, 1980 to present. Concerns have recently blossomed during corporate downsizings and restructuring. Also, drug incidents in the workplace have spotlighted concern about 24-hour work.

[12]The idea of "technical empowerment" is a concept that we have honed in our safety programs during the nineties, due to disasters that have occurred in oil, chemicals, transportation, and space. It is fairly easy for management to switch into *autopilot thinking* and forget the high potential of contemporary manufacturing and transportation equipment that our *new kids on the block* are operating.

# CHAPTER TWO

## Two Views of the Shiftworker

### OBJECTIVES OF CHAPTER TWO

**DARKSIDE AREA:** You will be able to discuss the Burnout Model and the "Double H Pattern" contributing to Set-Ups that all too many shiftworkers fall into and be able to recognize them in your workplace if they are occurring. You will be able to recognize common ailments that are typical of some individuals on rotating shifts.

**BRIGHTSIDE AREA:** You will be able to describe the achievement model of shiftworkers. You will be able to recognize employees who are enjoying their extra-achievement and just as importantly, those who are acting as positive role models for others.

By the time you have finished this Chapter, you will have developed an understanding of how the "self-fulfilling prophecy" affects shiftworker performance, health, and lifestyle through two extremely different views. From this understanding, you will be able to promote the high performance or "brightside" view of your 24-hour performers ... to the performer's advantage, the company's advantage, and to your own advantage as a practicing manager or trainer.

These abilities will assist you in supervising, promoting, and training your employees. You will also make better decisions on whether to send certain 24-hour employees to EAP (employee assistance) and other assistance.

### TRAINING — DARKSIDE OR BRIGHTSIDE?

We originally began studying 24-hour safety and performance as a corporate training challenge. Gene Lowe, a Dow U.S.A., Texas Operations training manager, had come to us with a shiftworker training need. He had a deep concern that shiftworker training was his area that could use the most improvement. Due to our past successes in producing lively "team-based" and behavioral programs for Dow, he and his management "co-opted" with us to look closely into the shiftworker job and develop training to fill the need.

Gene was especially concerned about his new-hire trainees who were asking for advice on how to start their 24-hour careers the best possible way.

An Unsettling Picture

As we studied this manager's existing shiftwork training programs on the shelf, we began to focus on a very unsettling scene that was occurring around the corporate community. New 24-hour employees were being exposed to commercial clinical-appearing programs that viewed them as workers with a dark problem needing treatment or prevention. They were viewed by some sleep researchers and psychologists as workers on the darkside, tolerating their abnormal job for extra bucks and a few long "breaks" each year. We definitely needed to create some training dynamite for Gene's company and lots of it. Fortunately, as you will see in this manual, we achieved just that.

Having 17 years shiftwork experience behind Gene, 20 years behind Glenn[1] and 15 years of shiftwork family behind Peggy, the client's "training program" quickly edged us toward the darkside as well! Empowered by the progressive Dow management, we set out to find training solutions that would zip us past these "darkside" problems.[2] In a minute, we'll give you a glimpse of this darkside.

Our research lead us to experts around the world. Incredibly, as you will see, much of the shiftwork solution came from only sixty miles away from our offices at a trainer's dream - the NASA Johnson Space Center in Houston, Texas. As we talked with the Space Center flight specialists,[3] we began to break out of our mindset about shiftwork and shiftworkers.

**The Prerequisite to Management and Training**

We found that as a vital prerequisite to shiftwork management and training, you must understand the two prevailing views of the shiftworker and their effects. Understanding these views and patterns of behavior associated with each, will make or break you as a manager or trainer in 24-hour work environments.

### *The Vital Prerequisite!*

Understanding the two views of shiftworkers and patterns of behavior associated with each, will make or break you as a manager or trainer in 24-hour work environments.

## EXTRA-ACHIEVERS AND SHIFTWORK

To prepare you for the two prevailing views of the shiftworker, let's lay the foundation by looking at a very special shiftworker.

### Electricity and Excellence — the Original Extra-Achiever

Thomas Alva Edison gets much attention due to his inventions. Sleep consultants always mention Edison and invention of the light bulb. However, one facet of Edison has been ignored by them and by teachers of history ... This 24-hour performer began a new line of 24-hour extra-achievers.

Edison was the originator of contemporary extra-achievement. His invention plans often had a built-in double whammy. His plan to invent the electric light bulb was propelled by several motives. One major function of the electric light was to allow individuals to become more like his ideal human ... a human that would be productive and work around the clock!

Edison tried to make it appear that he had learned to work nearly without sleep; however, we know better. You probably already know that he actually slept 3 to 4 hours at a time and also cat-napped throughout the day and night.

So, Edison earns his spot with us as a shiftworker and a modern society 24-hour extra-achiever. Certainly, his dream of helping other achievers work continuously around a 24-hour clock came true.

After Edison's invention of the light bulb in 1883, things really began to move ... a huge number of front-line workers began moving into shiftwork jobs and it was usually their own idea. As a 24-hour performer, they could put in more hours, produce more products, and earn more pay than the dayworker!

You know the rest of the story ... how the industrial revolution led to 24-hour processes in manufacturing, oil, and chemicals. And how new technology led to longer airplane flights, 24-hour emergency rooms, 24-hour patrols, 24-hour response systems ... and the problems that continues to haunt us ... jet lag and fatigue on the job!

The idea of Edison's mentality leads us as managers and trainers to take one more step. If Edison was an extra-achiever, and if his way of expressing it was to work around the clock, what about our own modern shiftworkers? How did they get selected into that 24-hour work? Did they just stumble into it ... or was there some self-selection rhyme and rhythm? Didn't most of them have a touch of that virus, "extra-achievement?"

It is our best thinking that they did! Research in this area is emerging.[4,5,6]

To get into the idea of extra-achievement in contemporary shiftworkers and how it is often thwarted, let's begin the case of your fictional nephew.

### *The Story of Sam, the Neph*

#### **Intro: The Neph decided to make it!**

Sam was your favorite nephew. By the age of nineteen, he showed signs typical of your family ... energetic, task oriented, with a definite knack for technical projects. He attended two years of junior college and upon graduation began looking for a job. He began his search by making a thorough survey of family members, friends, and job counselors, asking about possible job openings around the area.

During the past year, he had kept an eye on the classified job ads and had responded to several. The "Neph's" problem was that he could find only minimum wage jobs that held no current or future answer to his dreamed-of lifestyle.

The last two summers, he had worked at OxJohn Chemical performing odd jobs, since you had helped get him into their summer college student program. As he performed his duties, it had dawned upon him that the plant operators were something of a special breed with special work hours and other benefits. And it seemed they had important jobs. One night, when he was asking you and Uncle Jack about the jobs in your plant, the two of you mentioned the very difficult jobs of the shiftwork operators and began joking about their excellent pay scale and they're "driving better vehicles than you, the two top plant engineers."

Guess what job the Neph quickly went after? Do you think that he scored above or below the imaginary line on "achievement?" We think that he scored above the line. Well, he definitely wanted to belong to that group of extra-achievers, the plant operators! The Neph was willing to work harder than average in order to reap extra benefits. Let's move on to how the story progressed.

**Chapter One: The Neph has it made!**

You had been thrilled that Sam got himself hired so quickly into your company, OxJohn Chemical. He had finished two years at the community college and decided "that was enough education." You were initially a little disappointed that he jumped so quickly at the job of shiftworking operator, since both you and his Uncle Jack were college graduate plant engineers getting maximum pay. However, within a few months, you found that the Neph's total package which included overtime, shift differential, meal money, and other benefits, wasn't that far (if any) behind yours at that stage of life! On top of that, his new wife, Fay, had found a day job to add to their income. So much for a college education, an engineering degree, and a day job, you thought!

Two years later, the situation was changing.

**Chapter Two: The Neph has peaked!**

You found out about it during the annual family Christmas Dinner which was being hosted at Uncle Jack's that year. After finishing your second "Super-Granny Turkey Plate," it was time to hear success tales from the past year. Winking at Jack and looking around the table, you began to ask the Neph a few questions about how it was going. Fay rolled her eyes as Sam got into the story of their great hunting year.

Neph: "Well, I guess things have peaked at work. My supervisor told me that if I wanted a promotion, I could have his job ten years from now ... when he retires! Anyway, I've taken up hunting in the Hill Country with Jim and Jeff. It's a five-hour drive, but, gosh with 4 days off in a row, it's not that bad. We have a really great time out there, even when we go just to clean up the camp! Besides, there isn't much to do around here."

Fay mumbles: "Yes, and the baby hasn't met Jim and Jeff yet!"

Neph boasts: "We tend to have a few too many drinks sometimes, especially if we're rained out and stay in the cabin. Anyhow, it beats listening to the baby crying ... and Fay ... on long breaks!"

Fay: "A few? His truck looked like a beer company recycling wagon when they came home this last time! I don't know ... maybe those trips will keep him from getting grumpier!"

Neph: "Well, uncles, I guess I'm made like you ... work hard, party hard and rest when you can."

Fay: "No ... he is 'work hard, party harder, and catch him if you can'!"

Everyone at the table laughed. However, you could feel a cold tension developing between Fay and Sam.

Just then, Fay jumped up to check on the baby who had begun crying in the living room and the Neph's yawn closed the conversation.

**Chapter Three: The Neph has lost it!**

The following Christmas Dinner at the Neph and Fay's had to be replanned at the last minute. The Neph had been dodging you at work since the beginning of winter. The second week of December, you found out why. Fay finally called the in-laws and in a strange metallic monotone had broken the news. She and the Neph would be separating as soon as they could work it out. She was moving out of town with the baby.

Christmas Dinner was quickly rerouted to Uncle Jack's. This year, the Neph came alone. In addition to everything else, he had been working a lot of overtime so you and Jack didn't pry too much. However, after the second Super-Granny Turkey Plate, he squirmed and gave a few answers.

Neph: "Well, 'Unks,' I guess you're wondering what's been happening with me."

You and Uncle Jack gave him a calm look but didn't answer.

Neph's voice cracks: "Actually, I don't know. Fay says that I've gotten impossible to live with. She hates my favorite things ... hunting and driving to the Hill Country with Jim and Jeff. She's <u>so</u> tired of it, tired of me being 'irritable' and 'gone all the time' and I'm tired of being wakened by the baby crying and her doggone nagging!!"

Uncle Jack: "But it seems like you're going to lose a lot. Not to mention your little family — you just recently bought that new boat."

Neph: "I've already lost it. I'm tired of trying to keep up with the bills. I need to be living with someone who understands shiftwork — who knows what it's really like ... someone who likes the outdoors and doesn't want to spend a fortune on everything. She's even been talking about having another baby! I'm ready to get off this rotating schedule, out of this marriage, and head for the hills!"

That year, the family wound up dinner quickly. "My old day job is looking a lot better all the time!" you thought.

## An All-American Story!

This three-chapter shiftwork tragedy is repeated hundreds of thousands of times around the country each year due to the self-fulfilling prophecy of the darkside. Even though our typical shiftworker begins as a bright-eyed extra-achiever, he/she often succumbs to the darkside influence — due to a belief system influenced by spouse, boss, peers ... and aggravated by jet lag. The darkside influence promotes early burnout due to the simple self-fulfilling prophecy that his/her job and lifestyle are hopeless or not going anywhere, and that he/she is helpless to do anything about it![7]

And the prophecy runs its course within three to seven years as we will see in a minute.

The darkside of shiftwork often develops as in the case of the Neph and we can't deny that fact. We're going to quickly review the darkside and its problems, look at the Burnout Model, then get back to where we really should be ... on the achievement side, promoting the alternative: Performance Tracking and Improvement — a safe, proactive employee life and work style.

## THE DARKSIDE AND SHIFTWORK

It is well known that a great percentage of shiftworkers, like the Neph, complain of far more health and social problems than day workers. Common complaints their doctors, counselors, and supervisors hear are in the areas of sleep, health, mood, family, financial condition, and general attitude.

## Darkside Complaints:

- Fitness
  Excessive alcohol, drugs, smoking, eating, "gut" problems
- Alertness and Mood
  Irritability, malaise, depression
- Sleep
  Insomnia, fatigue
- Time-Off
  Family problems - Out of mainstream activities
- General Attitude Problems
  Capped, "in a rut" - Not getting much out of life
  Overloaded, Grumpy!!

### Darkside Alcohol, Drugs, Smoking, Eating Excesses

We know that shiftworkers often go to excess with habits. They definitely rack up a higher incidence rate of drug and alcohol abuse than daytime workers.[8]

Drug, alcohol, smoking, and overeating abuse tends to correlate with just about any high stress job assignment. Shiftwork is no different. We look at this abuse as reactions — coping mechanisms — to the added stress of shiftwork — coupled with the darkside prophecy.

### Darkside Gut Problems

A large number of shiftworkers report some form of "gut trouble."[9] Much of this is caused by long lengths of jet lag ... internal clock upsets, eating foods into the midnight hours. Many rotating shift employees often cannot tolerate food after 1 or 2 a.m. Glenn recalls stories of existing on a "very bland diet" for a year in his late twenties even though he loved the 24-hour lifestyle. The exact problem causing the abdominal cramps was never quite pinpointed by three very puzzled and well-paid physicians, we think we may have figured it out!

### Darkside Irritability, Malaise and Depression

Shiftworkers are known to be often irritable, grumpy and even depressive.[10] One research report states that European shiftworkers frequently complained of disturbances in their moods. Depending on the location, between 36 and 58% mentioned problems such as increased "irritability, nervousness, and bad temper." Most complaints were made by those working a three-shift rotating schedule, and by those who felt that they were hopelessly stuck in a permanent life of shiftwork.[11]

### Darkside Insomnia and Fatigue

Sixty percent of our 24-hour performers complain of sleep problems. This is well above the 20 percent of dayworkers.[12]

**VALIDITY CHECK:** Next time you visit your local medical doctor, ask if there have been any local shiftworker complaints. You may be surprised at his/her candid answers and the number of individuals affected by fatigue, gastrointestinal disorders, and emotional highs and lows.

### Darkside Family and Community Problems

Many 24-hour personnel become irritable around the family, seldom seen, and out of synch. If they undergo the reaction and burnout phase, we often hear statements such as the following:

## From the Family

- We always walk on eggshells when he/she's home.
- He/she's *always* grumpy.
- He/she used to be grumpy just on graveyards … NOW IT's ALL THE TIME!

**OR …**

## From the "Shiftworker"

- I don't feel like going with you.
- Go ahead and I'll stay at home.

**AND FINALLY ...**

### Eventually from Both

- I don't know any of YOUR friends!
- I don't want to know any of YOUR friends!
- I wouldn't like them EVEN if I knew them!!

Rotating shifts can be especially tough for younger, educated employees. All in all, younger employees want to concentrate on maintaining contacts and making new friendships. He/she is also interested in spending plenty of time with the new spouse and small children in the family.[13]

Actually, the greatest dissatisfaction with shiftwork has been found among young single workers and young married workers with small children.[14] Rotating and evening schedules collide with the social schedules of spouse and school children ... and with schedules of newborns who fit into a special class of shiftworker![15]

Community activities may not fare any better. Friends and community groups routinely report that they have a hard time tracking shiftworkers' rotating shift cycles. Shiftworkers eventually lose out on social visits and meeting invitations because of the endless rotating.[16] To the outsider, it appears that the shiftworker is constantly either working or sleeping, making it difficult to place the social phone call. In the F.A.S.T. Tracking section of this book, some solutions are presented to address these problems.

Researcher Donald Tepas has found evidence of high shiftworker divorce rates — compared to daytime employees.[17] We've seen entire departments and locations of 100-plus employees go through two-year "divorce epidemics." Crews brag on their second and third waves of divorces. This doesn't help overall morale, to say the least.

## Living with Shiftworkers

We get special views of the shiftworker from those that live with them. The following is an interesting report given by Charlie Kilsby, who was raised in the home of a chemical plant shiftworker during the forties and fifties:

### *Son of a Shiftworker*

"We lived in a wood frame house that was on blocks. It was like a drum. You had to walk through it very quietly. I called it walking like an Indian. You put your foot down on the floor without slapping your heel down, always on your toes.

"During graveyards, play was kept as far away from the house as we could. Laughing and giggling was not allowed inside the house. I break up and laugh sometimes about some of the things that happened, but now it's funny to me how we managed to get along and how tough it was.

"Shiftwork to a kid was like finding a toy that didn't squeak. You never played with anything that jingled or jangled. You learned to play games that were quiet. You had no video games in those days by any means. Just tic-tac-toe, and sometimes hopscotch. I just can't see Dad being able to stand the sound of video games. That, for sure, would have made him come unglued!

"It was hard for us children to sit down and relax around him at times. Things weren't all that bad, and they weren't all that good, either. Dad's temperament made things hard. To walk through the house, to slam a screen door . . . suddenly the old screen door would slam, and your heart would almost freeze. If it slammed and you woke up dad, you were in trouble!"[18]

## Darkside Job Attitude

On top of the physical and social issues, the darkside shows up on the job. Many 24-hour people complain of being capped, as Charlie Kilsby puts it.

He has heard this from thousands of shiftworkers in his South Texas barber shop. "They're missing the mainstream path and feel that they are not eligible for the same promotions as day people. This capped group often includes shift foremen."

### Darkside Team Morale Potential

Coupled with one or more of the reported problems, the general attitude and morale of the shiftworker and even the entire shiftwork team can become darksided and terrible. We hope this has not developed within your work teams, nevertheless the potential is there.

**When darkside group morale develops, nothing seems to please the shiftworkers or the supervisor.** Some refer to this as group-think. As soon as one employee problem is resolved, another crops up. Fixes such as new schedules, benefits, and engineering changes are quickly swallowed up by more problems (by both shiftworkers and supervisors).

As management begins to search for an "Iacocca of employee relations," the scene becomes consultant's fodder, perpetuating a never-ending trail of invoices for services. This is also a supervisor's nightmare. It's much safer and more productive to make an early push to develop the brightside potential of the 24-hour employee than to have this emerge in the entire group! Nevertheless, if this is your case, you may have some success championing a front-line driven performance improvement process.

## The Burnout Model

As already mentioned, the many complaints and problems of some shiftworkers quickly edged us toward the darkside view. The situation became one of not seeing the forest because of the trees.

Could we find the forest? Certainly, all shiftworkers weren't lost souls. A goodly number were doing really well.

We began to see 24-hour workers through two models — Burnout and Achievement and our perceptual clouds were lifted! The majority of shiftworkers fit into one of these two models. Over time, darkside influences lead to burnout and brightside influences lead to achievement — it's as simple as that! To see shiftworkers, we must view them over time! Here are the three phases of the Burnout Model:

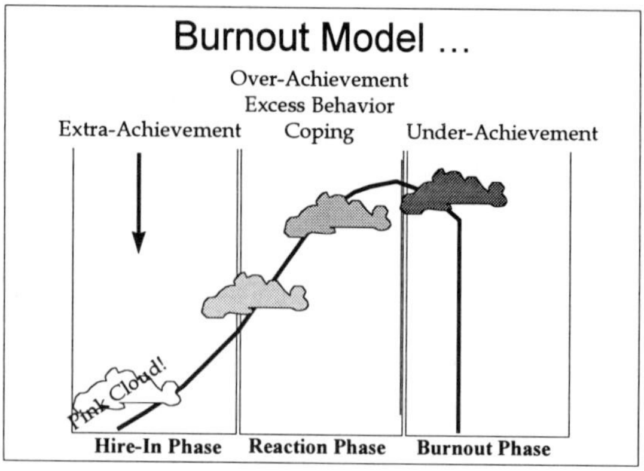


*Burnout Phases*

· **HIRE-IN PHASE:** During the initial hire-in period, our enthusiastic employee tries for extra-achievements that are available through shiftwork. Extra work, overtime and holiday pay are some of those extras. During this pink cloud — every thing is rosy period, the employee also receives several incremental salary increases. Due to the increased income, new time-off achievements are also begun.

· **REACTION PHASE**: During the reaction phase, we see a collage of important attitude changes and early reactions to stress. The extra-achievements on the job often prove short-term and their benefits or payoffs are soon forgotten. The individual begins to feel "in a rut" and hopeless about ever attaining these achievements. The employee suspects that he has been "capped"[19] and has become committed to a lifetime of continual shift rotation, minimal advancement, and below-par social activities. He also feels helpless to do anything about it — a dark cloud encloses his world.

Stresses from the job, jet lag, economics and family lead to distress reactions. Extra-achievement becomes focused more and more on non-work activities: sports, hobbies, hunting trips, vacations, home improvements. The overtime and holiday pay becomes necessary for everyday subsistence rather than the extras enjoyed during the hire-in phase, i.e., the canoe, the camper, the new house.

One or more excessive personal behaviors, as a way of coping with the stress, begins to emerge: excessive eating, drinking, smoking, sleeping or "staying out." What was once an activity enjoyed as entertainment, now becomes necessary to purge the stress and fatigue.

**BURNOUT PHASE:** During burnout, the physical and mental consequences of the initial stress plus the distressful coping behavior adds up to a crisis level. By now, the individual is noticeably moody or irritable, **even while working the day shift.** Feelings of achievement become muted. Close relationships have been lost with the spouse or old school friends due to "unpredictable schedules." New close relations become sparse and difficult to acquire. Perhaps the only friends the shiftworker has are co-workers on the same schedule. If there is a group-think of poor attitude, the problem may be compounded and symptoms severe.

The individual now feels permanently capped in his/her job, workstyle and lifestyle. The Double H Pattern of hopelessness and helplessness has settled in.

### Double H Pattern — A Tumbling Act

The shiftworker, in the third phase of the burnout model develops the Double "H" Pattern of helplessness and hopelessness. There are overwhelming feelings that life (day life — the real world) is passing by. If this develops, it will likely occur between the 3rd and 10th year of shiftwork.[20]

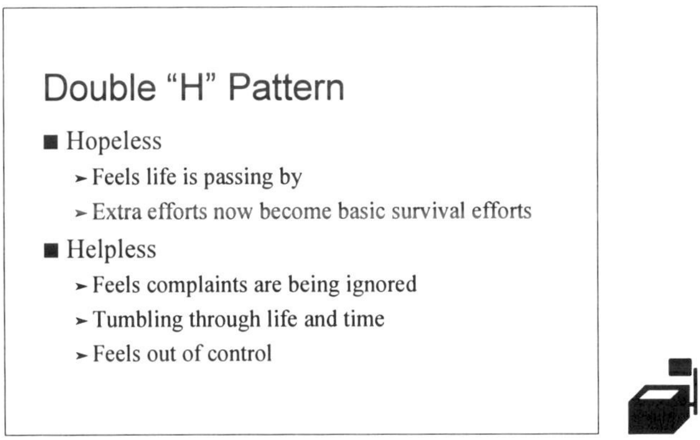


Even though the individual continues to put forth the extra effort of rotating and overtime, it is now necessary for lifestyle survival and a psychological subsistence mentality sets in.[21] In the Double "H" victim's mind, extra efforts both on and off the job lead nowhere; work and personal complaints fall on deaf ears; and all major purchases and decisions turn out wrong in the long run — missing their original good intentions.

Many attendees at conferences where Peggy speaks often comment that these complaints are not ONLY related to shiftworkers, but dayworkers as well. She reminds them that the reason we are focused on these reactions in shiftworkers is the Technical Empowerment and catastrophic safety potential. It is important that day working managers and supervisors pay attention to these signals as well since the alertness and decision-making of key personnel can affect the entire organization. *(See Challenger Incident, Roger's Report).*

A personal tumbling act, suspiciously mirroring the effects of the eight-hour cycle of evenings-days-graveyards and permanent jet-lag in an unending-uncontrollable rotating shift has been set into motion. We don't believe that this is the safest, most productive lifestyle for a productive employee on a critical job!

**NOTE:** These burnout symptoms look very similar to medical depression, and many shiftworkers have turned to their family physician for depression treatment. Unfortunately, a few have substituted "medical treatment" for their own proactive changes — fitness or diet modifications, sleep strategies, motivational time-off activities, or even requests for temporary day work.

To assure proper diagnosis and not a band-aid fix, be sure the medical advisor is thoroughly informed as to the employee's shift schedule, sleep habits, exercise and diet habits. This is not meant as medical advice, but as safety advice. If you are unable to modify your levels of alertness, or see negative changes in your behavior, seek help or contact your local physician, or if you are a manager, refer the employee to the proper Employee Assistance Program administrator for immediate help.

## BRIGHTSIDE VIEW OF THE SHIFTWORKER

### Successful Extra-Achievers

Many shiftworkers do not reach the Double H stage. In fact, they thrive on the 24-hour style — both on and off the job. This is more of what Edison had in mind!

### Successful Lives

We see 24-hour workers in very important jobs who can succeed in those jobs and their personal lives. Many nuclear plant operators and crack emergency response team members wouldn't trade their jobs or shift schedules for anything! And we find the same in chemicals, oil, mining, pipeline operations, law enforcement, manufacturing, and transportation.

## Successful Careers

Operators who eventually become Plant Superintendents or Technicians who develop thriving farm and retail businesses on the side are continuing their extra-achievements. So are those shiftworkers who continue their education and add to their job abilities throughout their careers.

## The Extra-Achievement Model

Individuals who continue achieving and improving fit into our Extra-Achievement Model. This model shows their success path over time. These employees do whatever is necessary to continue their extra-achievement tendency.

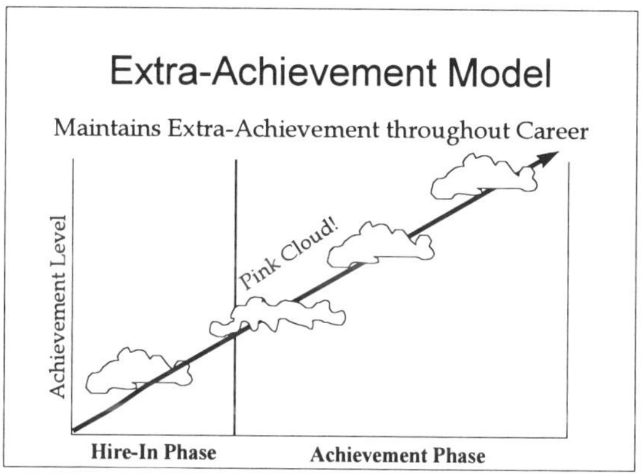


## Phases

- **HIRE-IN PHASE:** The initial hire-in phase looks the same as in the burnout model. However, a **healthy brightside, positive exposure continues** during the time our enthusiastic employee tries for extra-achievements. This positive image — pink cloud — may come from family, friends, the work team or supervision. Extra work, overtime and holiday work are enjoyed and the income is used for extra-achievements. As in the other model, the employee receives several incremental salary increases during the first year or two.

· **ACHIEVEMENT MAINTENANCE PHASE**: Rather than succumbing to the added stress of shiftwork, the employee takes proactive measures to continue the extra-achievements. These measures can be categorized in the areas of fitness, alertness, sleep, and time off — as in our F.A.S.T. Tracking theme. The individual keeps moderation in mind when signing up for additional overtime, making purchases, or enjoying a social life.

Rather than developing excessive reactions which lead to the Double H Pattern, the employee learns to handle this new lifestyle, takes control and excels. He/she even thrives on shiftwork until retirement or until taking a day job such as a Supervisor, Safety Specialist or Lab Technician.

## Origin of the Views of Shiftworkers

We found that employees (managers, shiftworkers, executives, etc.) and their families tend to view the 24-hour work style and lifestyle in one of two lights. It's either seen as an extremely good opportunity, promoting individualism and achievement, and is valued, or the opposite — it's seen as off-beat, hopeless, holding very few opportunities, and undesirable.

The achievement view comes from experts in success ... experienced 24-hour people that have worked at optimizing their career and family life while holding an around-the-clock job and the managers who have worked with them.

The strong negative view comes from real people who have not been able or have not tried to optimize their job performance and lifestyle. Many people are just not cut out for shiftwork physically or psychologically.

### SUMMARY — UNDERSTANDING THE TWO VIEWS

The dark and bright views of shiftworkers act as self-fulfilling prophecies that lead to the two shiftwork models: Burnout and Achievement. The majority of your shiftworkers will fit into one of these models.

Burnout leads to the Double H Pattern: Helplessness and Hopelessness. These twins of failure contribute to Set-Ups that lead to human error and they contribute to excesses and non-achievement.

Achievement, in the Achievement Model, becomes a career-long event that is maintained — by continuously working on four areas: fitness, alertness, sleep, and time off. Like continuous improvement on the job, individual performance for the shiftworker in a critical job and 24-hour lifestyle requires constant attention.

## CHAPTER TWO ENDNOTES

[1]Glenn remembers his first real shiftwork. While in the Marine Corps on two years of security duty, he was given the *opportunity* of working 24 hours and then spending 48 off. He'll never quit mentioning those stiff "Navy horsecock" sandwiches that he would pick up at midnight after a brief nap to fuel the last third of his 24-hour "day" and patrol a dark Nevada desert until 7 a.m.

Later, he would work four years of straight midnights as assistant shift supervisor of a hospital while a full-time day student. Next came 10 years as a rotating plant technician at Dow Chemical and then the over-achiever wrap-up ... a wholloping four years of 12-hour daywork as a supervisor (begun every morning at 5 a.m.) followed by commuting 60 miles to night school at 5 p.m. and home again around 1 a.m.!

[2]The Dow Chemical Company has been considered a leader in industrial safety for a number of years. Recently, it has emerged in the training area as well.

[3]The all-important Flight Activities Office at the Johnson Space Center, proved to be an invaluable source of information and insight. NASA exhibited the dynamic ability of the space program team to continuously learn and improve in the human factors area.

[4]Economists Jeff Biddle and Daniel Hamermesh of Michigan State University have studied economics and sleep (not shiftwork directly).

They find that *Americans sleep significantly less when there is money to be made.* Especially American men, who often use their sleep time to make extra money. From: Sleep? Why? There's no money in it, by Peter Passell, *New York Times*, August 2, 1989.

[5]Toufexis unknowingly roamed onto our turf in Time Magazine: "It was civilization that created the dilemma of sleep loss. The sun presumably dictated the habits of ancient people: when it was up they were awake, and when it went down they slept. Maybe when the moon was full they stayed up a bit later. The discovery of fire probably allowed the first change in that pattern. *As flames lit the dark, surely some adventurous souls delayed bedtime. But sweeping change came only a century ago with the introduction of the light bulb.* Thomas Edison's glowing invention permitted cheap, safe and efficient illumination throughout the darkest nights..." From: Drowsy Americans, Anastasia Toufexis, *Time*, December 17, 1990.

[6]Dr. Jamal has independently made relevant headway into the makeup of individuals who often work night and day — moonlighters. He highlights two competing schools of thought:

(1) They are a special breed, with more energy, as well as higher social and economic expectations than the average person — OR —

(2) They are, in general, economically squeezed and socially deprived. *In his study, moonlighters working in five plants in two continuous process industries scored very high in the achievement area.* From: Is Moonlighting Mired in Myth? by Muhammad Jamal, *Personnel*, May 1988.

[7]During earlier history, the prophecy was backed up by more realistic physical and economic barriers. Shiftworkers were less educated, there was less worker involvement and fewer achievements available. Twenty-four-hour jobs were less critical, plants and departments were smaller and featured less advancement potential.

[8]Shiftworkers suffer a higher incidence of drug and alcohol abuse than daytime workers. See: Coping With Your Body Clock, *Occupational Hazards*, November, 1991. Glenn McBride and Peggy Westfall's interviews with employees in chemical plants and refineries support this comparison. See also the "over-achievement" phenomenon described later in our Burnout Model.

[9]Dr. Timothy Monk reports that he knows of very few shiftworkers who don't suffer in one form or another of gut trouble. This ranges from poor appetite to excessive appetite, to diarrhea, to constipation. See: *How to Make Shiftwork Safe and Productive* by Timothy H. Monk, Ph.D., published by American Society of Safety Engineers, 1800 East Oakton Street, Des Plaines, IL. 1988.

[10]We tend to think of "depression" as something that "little old ladies" and laid-off employees experience. However, it can hit the "able-bodied seaman" just as well.

[11]Reported in: *Shiftwork — the Social, Psychological, and Physical Consequences* by Paul E. Mott, Floyd C. Mann, Quin McLoughlin, and Donald P. Warwick, Ann Arbor, University of Michigan Press, 1965.

[12]From: *Losing Sleep — How your sleeping habits affect your life* by Lidia Dotto, Morrow, NY, 1990.

[13]From: *Shiftwork — the Social, Psychological, and Physical Consequences* by Paul E. Mott, Floyd C. Mann, Quin McLoughlin, and Donald P. Warwick, Ann Arbor, University of Michigan Press, 1965.

[14]See: *Shiftwork — the Social, Psychological, and Physical Consequences* by Paul E. Mott, Floyd C. Mann, Quin McLoughlin, and Donald P. Warwick, Ann Arbor, University of Michigan Press, 1965.

[15]Sleep researchers tell us that newborns do not have their internal clock set until some time has passed!

[16]An excellent academic write-up on the social effects of rotating is found in: *Shiftwork — the Social, Psychological, and Physical Consequences* by Paul E. Mott, Floyd C. Mann, Quin McLoughlin, and Donald P. Warwick, Ann Arbor, University of Michigan Press, 1965.

[17]In one study of 1490 workers, Donald Tepas and colleagues found an almost 50 percent increase in divorces of night workers compared with day workers. From: Flexitime, compressed workweeks and other alternative work schedules by Donald I. Tepas in *Hours of Work — Temporal Factors in Work Scheduling,* edited by Folkard, S. and Monk, T. H., John Wiley & Sons, 1985.

[18]Excerpts from: A two-hour taped study of living with a shiftworker by Charlie Kilsby, 1991.

[19]"Capped" is a term often used by Charlie Kilsby to describe the feeling of the plateaued worker.

[20]The *Burnout Cycle* and *The Double H Pattern* are concepts that summarize the undeniable data from multiple researchers, practitioners, and a multitude of 24-hour workers. Mott et al. have made wonderful empirical progress in this area. See: *Shiftwork — the Social, Psychological, and Physical Consequences* by Paul E. Mott, Floyd C. Mann, Quin McLoughlin, and Donald P. Warwick, Ann Arbor, University of Michigan Press, 1965.

[21]Timothy Monk most aptly terms this the "treadmill." From: *How to Make Shiftwork Safe and Productive* by Timothy H. Monk, Ph.D., published by American Society of Safety Engineers, 1800 East Oakton Street, Des Plaines, IL, 1988.

# CHAPTER THREE

## Shiftwork and Safety

### OBJECTIVES OF CHAPTER THREE

By the time you have finished this Chapter, you will be able to give an overview of the impact of shiftwork on safety. You will be able to use this information (coupled with that of "Section 1 and 2") to make a pitch for your own shiftwork safety, for a new schedule or for a new training program.

You will be able to state shiftwork-related safety problems and be able to identify potential industrial incidents where shiftwork will be a major factor.

You will be able to describe shiftwork fatigue, jet lag, and Danger Zones, and the influence these factors have on Shiftwork Slips.

### COMMUNICATING AND SELLING SAFETY NEEDS

Our purpose in this chapter is to make you aware of shiftwork safety problems and to help you communicate and sell your safety needs to both upper management and employees.

Although we realize that some industries are making valiant attempts to overcome these problems through engineering, policies, and regulations, we will not survey those efforts in this section. In the later sections on Scheduling and F.A.S.T. Tracking Techniques, we will give our specific suggestions and the proven techniques that may help overcome the problems.

## Human Error in Shiftwork

Scientific studies of industrial and transportation incidents typically find that about 90 percent are caused by direct human error (Slips) or indirect error (Set-Ups).[1]

A very highly-respected sleep expert, Dr. David Dinges of the University of Pennsylvania, has worked with the 90 percent human error figure. He believes that the Set-Up of inadequate sleep is a major factor in human error. It is at least as responsible as drugs, alcohol and equipment failure.[2]

Also: Employees are more likely to commit this human error between midnight and 6 a.m., even when they have slept well during the day![3]

### Shiftworkers Have More than Their Share!

The shiftwork and human error problem is becoming better known at the national level. At a meeting of the American Association for the Advancement of Science, Charles Czeisler stated that police officers, nuclear power plant operators, medical interns and many other night workers perform more poorly and are involved in more on the job accidents than their daytime counterparts. He mentioned that rotating shiftworkers were at the top of this list due to their internal clock disruptions.[4]

**Most Admit to Falling Asleep**

And it isn't just Set-Ups and Fumbling Physical Slips[5] that are showing up as the cause of incidents. Seventy to 80 percent of individuals who work night shiftwork admit to falling asleep at work (committing Attention Slips)[6], according to Dr. Martin Moore-Ede, founder of the Institute for Circadian Physiology and associate professor at Harvard Medical School.[7]

In an industrial-related study, Dr. Charles Czeisler reported that his researchers collected data from 4,000 shiftworkers and found that 56% of these reported nodding off at least once per week. We classify nodding off as an Attention Slip."[8]

**WORLD CLASS INCIDENTS**

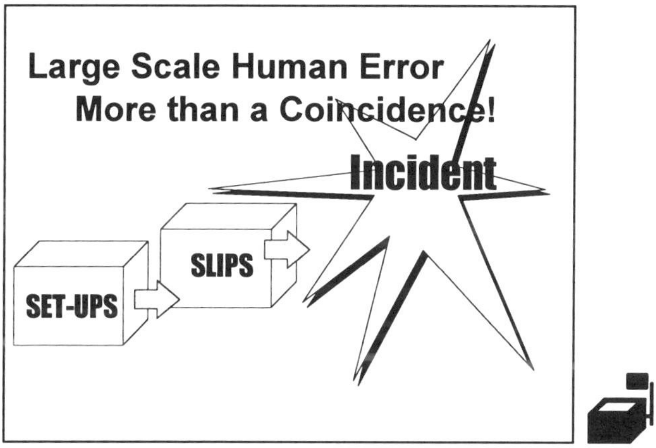


Verifying the extra shiftwork-related human error, Scripps Clinic and Research Foundation director, Dr. Merrill Mitler, reported finding "a link between the brain's sleep processes and such world scale disasters as Chernobyl and the Challenger explosion." He stated that it "appears to be more than coincidental that four major nuclear accidents occurred in the post-midnight hours (the nightly Danger Zone)." Last-minute decisions on the Challenger launching also were made before sunrise on that fateful morning.

A disproportionate number of destructive human errors (Slips) have occurred during the same time of the day that the brain has a tendency to fall asleep (the early morning "Danger Zone" as we label it in this book) Mitler's panel concluded. Much of modern life, with its emphasis on 24-hour operations, "is basically asking too much of the human body."[9]

Research coming from the "internal biological clock" angle shows that employees working long, irregular hours are definitely less attentive, think and remember less clearly, and have more accidents than dayworkers.[10]

### *Chernobyl and its Costs*

Chernobyl is a well-known shiftwork catastrophe that occurred in the early morning hours. This case of immediate human error coupled with planning and engineering Set-Ups continues to spin off disastrous consequences. The Supreme Soviet legislature adopted a $26 billion program to move 200,000 additional residents from the contaminated area. Officially, there were 31 fatalities. However, unofficial reports estimate a minimum of 250 deaths. The spread of Iodine 131 quickly collected in the thyroids of thousands of radiation victims.[11] The fallout from this incident will continue into the next century and for generations.

### *Other Late Night Incidents*

Other late night incidents often associated with shiftwork Set-Ups and Slips include the Bhopal Gas Release, the Three-Mile Island nuclear incident, and the Exxon Valdez oil spill.

## Fatigue on the Job

Shiftwork often goes along with the Set-Ups of irregular hours, rotating schedules, and excessive overtime. The cost of the resulting reduced job alertness has been estimated at $70 billion a year.[12] We believe that much of this reduced alertness shows up as fatigue and jet lag caused by the Set-Ups.

### *Congress Investigates Effects of Sleep on Safety*

Legislatures have gotten into the act. In November, 1991, the National Commission on Sleep Disorder Research, mandated by Congress, completed their 20-month study on inadequate sleep in the United States.

The commission concluded that insufficient sleep is an unseen killer. Merrill Mitler reports that millions of accidents each year are caused by drivers and workers who try to accomplish too much with too little sleep [13]

Mitler writes that these findings quickly followed a September 1991 report by the Congressional Office of Technology Assessment. This report said that the "biological cycles that affect sleep, fatigue and human error in the workplace have been underestimated or ignored by federal regulators, industrial management, and labor leaders."

Mitler says that recent research shows that individuals are more likely to make errors between midnight and 6 a.m. even when they have slept seven or eight hours during that day. (And they are EVEN MORE LIKELY to commit errors when they have not slept well during the previous 24 hours.) Gene Lowe, past training manager at Dow Chemical, labeled the dangerous hours as "the Danger Zones." We liked his label and have incorporated it into our work. There are actually two Danger Zones each 24 hours.

**The Danger Zones**

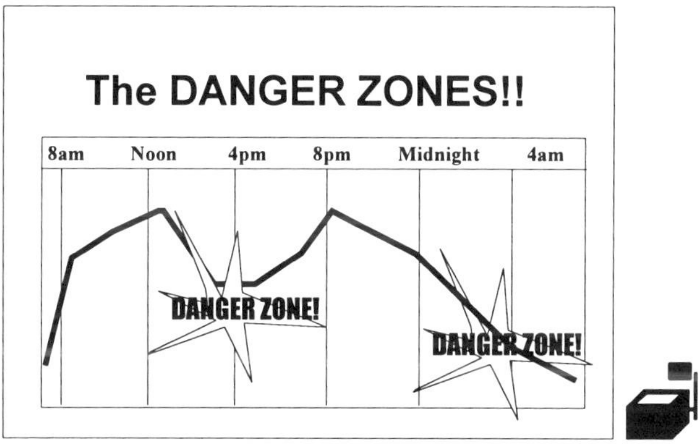


The most ominous Danger Zone, that Mitler described, occurs after midnight.

The Danger Zone "alertness dips" follow the daily body temperature cycle each 24 hours. This "circadian cycle" is usually aligned with real time (even for many long-term "night workers").[14]

Safety researchers consistently find a higher statistical rate of incidents occurring during the early morning hours. For example, between 4 a.m. and 6 a.m., the rate of fatigue-related incidents for single trucks is 10 times as high as the rate during the day.[15]

Also, an incredible percentage of major catastrophes have occurred around the Danger Zone time period! The Exxon Valdez touched ground just a few minutes after midnight. This is the same hour of the major gas release by the Union Carbide Plant in Bhopal.

The Danger Zones have been verified by many expert studies, including that of Merrill Mitler. Researcher Michael Coplen also studies this phenomenon. He cites a study of train engineers in Sweden: 10 percent reported micro-sleep Slips (of 30 seconds to 3 minutes) nearly every night.[16] In the same article, he reports that a Japanese study of 198 near-miss train accidents reported that 82 percent occurred between midnight and 8 a.m.

Coplen mentions the Chernobyl incident which began at 1:35 a.m. and the Three-Mile Island incident which occurred at 4 a.m.

**Aircrews in the Danger Zone**

NASA statistical studies have also verified the early morning Danger Zone. In ASRS database of aviation near-misses fatigue-related incidents tend to occur more frequently between midnight and 6 a.m. Curt Graeber states that these reported near-misses (Slips that did not become disasters) by crew, pilots and air controllers tend to be substantial and not just minor errors in airmanship.[17] He also states that variations in aviator flying performance can be predicted and observed according to the time of day.

Aircrews often get very drowsy during the Danger Zone. Graeber tells of studies that monitored their performance after midnight. While the airplanes were cruising on autopilot between 0400 and 0600, the crew would display "extreme drowsiness."

## SHIFTWORKER JET LAG

### Definition of Jet lag

Jet lag is the off-feeling one experiences when the internal clock gets off synch with real time for some reason. This term was first made popular by pilots and passengers on jet flights where it was produced by flying rapidly across several time zones.

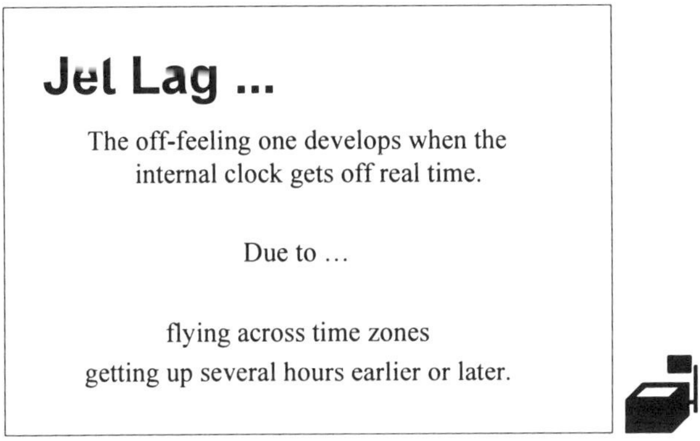


Our 24-hour people can develop a perfect case of "lag" by getting up and going to work several hours earlier or later than the day before.[18] When the shiftworker experiences jet lag, he/she feels like the real world is eating and sleeping at the wrong time![19]

## Jet Lag Affects Job Performance

Sports teams and Olympic athletes have learned to combat jet lag. They "fly in early and catch up with the clock" before they begin a game or event. Also, some international companies won't allow managers to sign a contract until 24 hours after crossing an ocean.[20] We wonder if those same companies have learned to pay as much attention to their other time travelers ... their local shiftworkers in critical jobs![21]

> **With Jet Lag ...**
>
> Athletes can't perform!
>
> Managers can't sign!
>
> Can our 24-hour employees function well?

Permanently "punch-drunk" is a very apt description that Timothy Monk gives to rotating shiftworkers when their internal clock gets out of phase with their work time for months or even years.[22] Is this really the best state for our employees in critical jobs making crucial life/death decisions?

## Where Jet Lag meets Fatigue: the Early Morning Shift Change

Communications set-ups can cause a fumbling or attention slip around daybreak when fatigued night employees hand over the night helm to a jet-lagged morning relief crew. This mindless relief situation can be fueled by the other set-ups of overwork, a rush, boredom, rotating hours, low quality off-time, poor fitness, poor eating habits, and lack of sleep or a combination of all these!

Psychologists tell us that employees are less likely to communicate well or intervene in situations such as this when they are experiencing the "vulnerable zone." Safety investigators are reporting on a fatal accident that may turn out to be a classical case of poor communications leading to an Attention Slip during early morning shift change:

**A Case of Forty-Three Screws**

A Continental Express commuter plane out of Houston on a circuit back from Laredo crashed mysteriously on September 11, 1991 in a cornfield in South Texas. The incident killed all 14 aboard. Investigators were soon reporting that the leading edge of the left horizontal stabilizer was apparently missing at the crash site. This horizontal stabilizer was "crucial" for control of the craft according to Embraer, the Brazilian manufacturer. If the stabilizer came off in flight, it could cause a sudden pitch violent enough to break off the left wing.

On the previous graveyard shift in Houston (September 10th), night crews had been assigned to replace the plane's right and left deicing boots. These rubber-coated strips fit behind the leading edge of both stabilizers. Therefore, the leading edges had to be temporarily removed in order for the new boots to be installed.

The two mechanics working on the right deicing boot completed their job according to the NTSB investigators. However, to help with the project, a supervisor had apparently begun work on the left side. The workers never finished the job on this side that the supervisor started. They didn't reattach the 43 screws that would hold down the left leading edge. (The aircraft's missing screws had been found in Houston by the investigators.)

The NTSB has stated that supervisors normally do not perform such work themselves. Also, the incomplete work on the left boot was not documented on the company work card. (The work on the right boot was recorded.)

For some reason, the plane left Houston with the log showing only one side was completed. Why did no one notice this?

When jet lag meets fatigue, anything can happen ... a storage tank can run over, a reactor can overpressure, an air craft can take off without all the screws.

***Reducing Slips with the Sun-Up Effect!***

On the other hand, adding some daylight during a crew's shift change can probably reduce the chance of error in many situations. James C. Miller, an expert in human vigilance and aviator performance, tells us that the chance of slips occurring around daybreak is much less likely if the employees have access to natural light. When he flew nights as a pilot, the cockpit crew always became alert and communicative at first light. He terms this the sun-up effect! Dr. Miller suggests that bright artificial light might be useful to reduce the risks of slips during critical work or shift change periods.[23]

***The Log Book at Shift Change***

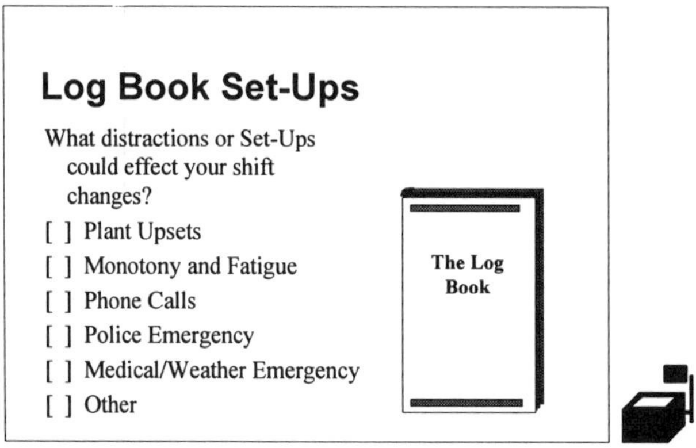


The log book often contains vital but silent information that is easily overlooked or forgotten during shift changes accompanied by shiftwork set-ups and distractions. Distractions can range from a chemical plant process upset to a medical or weather. The log book can hide vital information during this time, especially if an upset occurs during the shift change.

## SHIFTWORK SAFETY AND CRITICAL PLANT JOBS

### Nuclear Plants

#### *Asleep at the Helm*

U.S. nuclear plant safety has fared much better than Soviet nuclear safety. One reason is the strict U.S. safety and environmental regulations.

Another is the close monitoring and willingness of agencies to assess fines when these regulations are violated — averaging one fine per week since 1987.

In spite of this, intentionally snoozing nuclear plant operators have contributed their share to these fines. Adding more operators to the night shift as a cure to the problem in one plant only added to the sleepers that were caught a year later!

As reported in the Associated Press, the largest (Nuclear Regulatory Commission) fine of the early nineties was a $1.25 million penalty to the Philadelphia Electric Company (and also $500 to $1000 fines to 33 workers and supervisors). An internal informant, at great risk to his job, had told of nuclear power plant control room operators regularly sleeping while on the job.[24]

Such incidents teach us that sleeping on shift can occur in the most critical of jobs if we do not maintain the best of preventative procedures and countermeasures.

### Chemical Plants and Refineries

#### *Don't Wake 'Em Up! — Peer Pressure*

It isn't always okay to wake 'em up! One anonymous informant from the refinery industry tells of a new shiftworker's nightmare: When working nights, he felt he had to enter the refinery control room very quietly.

This was to prevent waking up older control board operators that might be napping. As the new guy, he certainly didn't want to get on the bad side of their group-think.

## Transportation — A History of Human Error and Fatigue

Transportation jobs are laden with night incident and examples. Many mysteries will never be solved. For instance, the Champlain riverboat ran ashore under full steam on a perfectly clear July night in 1875. No one has ever figured out why.[25]

### *Incident Coincidence: Stealth Fighters*

More recently, two Air Force Lockheed F-117A Stealth Fighters mysteriously crashed during the Danger Zone while on identical conditions and missions. This coincidence cannot be ignored. The Air Force cites pilot fatigue and "spatial disorientation" due to a rigorous night flying schedule as probable causes. We may never identify the final slip committed by the pilots. However, Lt. Col. John Miller, one of the three squadron commanders, had pre-warned in a July 10, 1986 memo that the extended hours were taking a toll on the pilots' performance. He mentioned detecting more and more instances of poor pilot judgment that were not occurring before the heavy flying.[26]

### *Transportation Star Trek in the 1990s*

A look at critical transportation jobs gives us an incredible vision of a transportation job world we're approaching. This is coming about through technical empowerment of the 24-hour individual. Imagine this scene occurring on a daily basis around the globe: Technically empowered 24-hour teams embarking on critical journeys in their multi-million dollar crafts (railcraft, aircraft, groundcraft, marinecraft, spacecraft) — with souls and hazardous payloads aboard at odd hours ... and with minimum supervision from home base.[27] Viewing it this way, we are quickly implementing our own exciting version of Star Trek!

**Transportation Star Trek**
**... in the 1990s and beyond**

Technically empowered 24-hour teams in multi-million dollar craft, embarking on critical journeys at all hours every day!

- With Passengers
- With Hazardous Payloads

As this scene suggests, 24-hour transportation jobs of the nineties deserve all of the shiftwork innovation that we can muster. Previous NTSB investigator John Lauber tells us that all-night air, train and truck operations have become facts of modern life. "That is not going to change so we must develop ways to help people deal with the unique biological and psychological demands of such operations."

### *Trains and Trucks*

Trains and trucks are land craft that have grown larger and more technically-empowered over the years. In many ways, the two types of craft have competed for hazardous materials payloads. The Congressional Office of Technology Assessment has stated that 62 percent of dangerous-material spills are caused by human error, not technical failures.[28]

### *Trains: Sleepy Engineer and Crew*

Shiftwork set-ups have led to many train incidents over the years. A sleepy engineer and crew were the probable cause of the 1988 head-to-head collision of two Conrail freight trains near Thompsontown, Pennsylvania according to NTSB investigators. This crash cost four lives and $6 million.[29]

### *Incident Coincidence: Engineer and Brakeman Asleep*

Coincidences and patterns of set-ups, slips and incidents occur in trains as in Stealth fighters. A Burlington Northern freight train struck another Burlington train head-on at 3:55 a.m. and five crew members were killed. Nine days later, another Burlington rear-ended *another* Burlington around 4:18 a.m. near Pedro, Wyoming and two were killed immediately. The National Transportation Safety Board investigated the accidents and determined that the employees' falling asleep (Attention Slips) was a probable cause in both incidents.[30]

### *Rails and Computer Control*

Rails are progressing with computer control. However, computers on the rail craft have not reduced the alertness needs of the crew. The engineer must continue to maintain an awareness of the track curvature, grade, and train length. Even with computer control, the human performer acts as the ultimate decision-maker on this critical job.

Rail officials feel strongly that it is the responsibility of the crew members to be well rested and in good shape on the job.

The rail employees we have contacted agree that certain set-ups causing fatigue need to be eliminated by rail shiftworkers. At the same time, they feel that action should be taken to promote better employee shift scheduling. Split schedules — where an employee works an 8-hour shift, attempts to sleep for 3 or 4 hours in less than ideal conditions, then works an additional shift — are the focus of ongoing safety studies.

### *Trucks*

The U.S. Dept. of Transportation reports that up to 200,000 traffic accidents each year may be sleep-related. They claim that 20 percent of all drivers have dozed off (micro-sleep slips) at least once while behind the wheel. A long-haul driver covering up to 4,000 miles in seven to ten days often averages only two to four hours of sleep a night.[31]

The Better Sleep Council has become concerned about the dozing statistics. They tell Americans that falling asleep at the wheel is second only to alcohol as a cause of U.S. accidents. They mention that drowsy vehicle drivers (including trucks, automobiles, etc.) kill as many as 6,500 per year.[32]

### *Truck Driver Fatigue*

A study of 182 fatal truck incidents in 8 states between October 1987 and September 1988 supports this. The National Transportation Safety Board found that impairment of the driver due to driver fatigue was a primary cause in 31.5% of the crashes.[33]

One-third of the one million truck drivers on the nation's highways may be too tired to stay awake while driving or to respond quickly enough to avert an accident.[34]

### *Truck Drivers and Drugs*

The same NTSB study found that one-third of the drivers killed in the crashes tested positive for alcohol, marijuana, cocaine, or over-the-counter stimulants. A full thirteen percent of the total tested positive for alcohol or marijuana.

**Truck Incidents**

| | |
|---|---|
| Fatigue-Caused: | 31.5% |
| Drivers on some drug: | 33% |
| On Alcohol or Marijuana: | 13% |

### *Truck Drivers and Micro-Sleep*

Many trucking accidents are due to simple nodding "micro-sleep."[35] With the number of truckers on the road increasing, individual incidents are too numerous to study in this book. By picking up your local morning newspaper, you will soon learn to instantly find sample incidents.

### *A Typical Truck Incident*

In today's paper, according to the Associated Press, between thirty and forty houses were evacuated west of McKinney, this morning. The evacuation was due to hydrofluoric acid escaping from a tank truck. The gas release was caused by an early morning wreck where one person died. The wreck happened about 5 a.m. (during the Danger Zone). It was a truck and car collision according to spokesperson Elaine Clayton, of the Texas Department of Public Safety.[36]

This accident brings up a major concern of all modes of transportation ... that of carrying hazardous materials through communities.

An insider with the rail industry reported to us that the community fear of chemical releases has prompted his crews to employ the dangerous and illegal practice of removing hazardous material placards from the rail cars while moving through certain towns.

### *Highest Trucker Incident Rates*

The hours of trucker incidents parallel other incidents around the country. According to a recent NTSB report, if you are a single truck driver driving between 4 a.m. and 6 a.m., you are ten times as likely to experience an accident.

### *Aviation — Congress Expands Aviation Research*

Congress has become very concerned about human factors in aviation. In September, 1989, Congress ordered aviation's FAA to expand its human factors research and to establish a human factors committee. The bill provided $25 million to study shiftwork-related items. Pilots (transportation shiftworkers) continue to cause the majority of airline accidents.[37]

Certainly, investment in aviation safety spins off into many areas. Since 1987, we have received a continuous benefit by applying the aviation findings to "industrial human factors" research. We credit aviation research for much of our improved insight into teamwork during emergencies, crew communications, the chain of human error, "autopilot behavior," and safety intervention.[38]

### *Asleep on Automatic Pilot*

Sleep experts do a good job of keeping us on the edge with two aviation stories:

In 1978, a Boeing 707 scheduled to land in L.A. right after midnight passed right over the airport at 32,000 feet and unaccountably headed out over the Pacific. Guided by automatic pilot, the aircraft was 100 miles out to sea before the air traffic controllers figured out how to sound an alarm in the cockpit. The entire crew had fallen asleep!

In another incident, a pilot flying a Honolulu-bound jet with 150 aboard, fell asleep at the stick when the aircraft was just 200 feet off the ground. The co-pilot realized what was happening in time to make a safe landing.[39]

***Technically Empowered Air Traffic Controllers***

Air Traffic Controllers have become nationally known due to publicity about their high stress jobs. The ASRS database of near misses includes errors committed by Controllers as well as cockpit crew members.

Austin Kiplinger predicts that Air Traffic Controllers will soon have "systems that can handle ten times today's traffic."[40] This is certainly a case of more technical empowerment of a job. Along with the increase in the number of aircraft in the sky and ground, there is technical empowerment — the aircraft will be holding more passengers and flying longer flights.

***Space — A New Round of Shiftworkers is Coming***

NASA will continue focusing on 24-hour work and human factors as the Space Station project progresses.[41]

Success in this latest round of shiftwork will determine the outcome of a vital space mission. Johnson Space Center personnel tell us that the space station management will be very much involved in this 24-hour work since they may work regular hours (on earth) AND make major decisions "after hours."[42] At the same time, crews will be multi-national. There will be many communications and scheduling lessons to be gleaned from this project.

***Kennedy Incidents***

The importance of shiftwork and overtime as safety issues is underscored by the incidents that have been reported at Kennedy Space Center in recent years. The Lockheed Space Operations Company Incident Error Review Board reviewed a total of 117 and 147 incidents during calendar years 1984 and 1985. Of these incidents, 51.3 and 56.5 percent were attributable to some type of human error, including procedural deviations, miscommunications and safety violations.

***Operator Fatigue on Launch Day!***

The following is from a U.S. government report of a Kennedy Space Center near-disaster:

"One potentially catastrophic human error occurred 4 minutes, 55 seconds before the scheduled launch of (shuttle) mission 61-C on January 6, 1986. According to a Lockheed Space Operations Company Incident Report, 18000 lbs. of liquid oxygen was inadvertently drained from the Shuttle external tank due to operator error. Fortunately, the liquid oxygen flow dropped the main engine inlet temperature below the acceptable limit causing a launch hold, but only 31 seconds before lift off. As the report states, 'Had the mission not been scrubbed, the ability of the Orbiter to reach defined orbit may have been significantly impacted."[43]

The report then mentions that operator fatigue was cited by Lockheed as a major factor in the incident. Launch control operators had misinterpreted system error messages resulting from a failed microswitch on a replenishment valve.[44]

To sum it up, operator fatigue had set in and a safety intervention 31 seconds before liftoff prevented catastrophe on this mission!

### *The Challenger Incident*

The Challenger incident is a case where effective safety intervention did not occur.

The Rogers Report on the disaster cited several possible factors contributing to the disaster, including excessive worker overtime. Overtime and fatigue in management were also cited, particularly for those involved in a telephone conference call between Marshall Space Flight Center personnel at Kennedy and Morton Thiokol employees on the Challenger booster/temperature concerns.[45,46]

In a human factors analysis of the Challenger incident, they mention that excessive overtime and fatigue of key individual personnel affect the safety of the entire operation. This is more important than the simple overtime average of the group! During the month of the disaster, several key individuals were essentially working two shifts. This type of data will certainly be identified more and more during incident investigations.

**Monitoring Overtime Set-ups**

Key Personnel

- Managers
- Technical

... as well as the group average

During our on-site shiftwork assessments, we review accident data and overtime records. Key personnel would include hands-on operators, engineering specialists, and decision-making management personnel.

Not only is overtime a key factor, but the ability of these key individuals to make decisions throughout a 24-hour clock is a major factor. Peggy makes herself not-so-popular with the management types when she suggests unannounced emergency response drills that involve all key personnel at 6 p.m., 11 p.m., 2 a.m. and 4 a.m.

**Shiftwork Safety and Critical Emergency Response Jobs**

Emergency response personnel are some of our most important shiftworkers. We wish we had space in this publication to review their many accomplishments and techniques that have gotten them to an extreme level of proficiency. A former Canadian Mounted Policeman, who trains emergency response professionals sums up the importance of training, performance, and alertness in this group: "When performing emergency procedures, there can be only 100 percent perfect performance. There is no room — no time for error!"[47]

We can add to this for shiftworkers in general: There is no room for set-ups that lead to errors! A special set-up we will cover in future publications is the scheduling set-ups in emergency response. Twenty-four hour shifts add unique alertness issues.

### *Medical*

The dedicated medical profession is an area of classical 24-hour worker abuse. The "workday" for interns and residents can stretch to longer than 24 hours and the "week" to 120 hours. Doctors on duty squeeze in brief naps when they can, but frequent interruptions from patients, pagers, and emergency phone calls — often leave them off-center and at a lower level of performance.[48]

Because of abuses highlighted in one case, New York state regulations have placed upper limits on the time interns and residents can remain on duty.[49]

### *Police*

Police departments and police officers are also notorious for abusing themselves by overworking and undersleeping.[50] Police researcher William Cunningham tells us that in addition to the many officers working shifts, about 30 percent of the nation's 500,000 police employees work extra (security type) jobs.

In Houston, the heat of moonlighting is really on ... Karen Roebuck reports that 85 percent of the Houston Police Department works extra jobs. However, thankfully there is a limit! " ... Under department rules, an officer can work up to 76 hours a week, including the time spent in court. One officer said that the department restricts their hours so they'll stay awake on their regular job."[51]

Amazingly, in 1990, 96.5 percent of the Houston police force's off-duty jobs were of police-type work. When you pass a Houston Police Officer, better check to see if his/her eyes are open!

In some cities, police officers very often cope with their schedule overload Set-Ups by parking the patrol car in a quiet spot and taking an unauthorized nap (sometimes called "copping Z's or cooping")[52]

In one survey of police officers (not Houston), 80 percent said they fall asleep once a week while on the night shift.[53]

## EXPERIMENTS IN SAFETY AND PERFORMANCE — DOING SOMETHING ABOUT IT!

### Reducing Fatigue through Napping

Researchers are concurring more than ever that planned napping is a good method of maintaining, even restoring performance in work crews.[54] This could be useful where a heavy workload spans several weeks, as Glenn experienced in plant "turn arounds" and South Pacific military games. As you will see under the "Napping Procedure," the nap cannot be used to catch up on sleep. It must be used as a countermeasure, proactively, before the sleep deficit shows.

**Napping Studies and Curt Graeber**

Curt Graeber has studied flight crew performance and fatigue over the years. Long-haul crews have been a special problem for the aviation industry, especially those crews that are overworked and under-rested. "Long hauls" can stretch the crew to its limit — Buenos Aires to London, or San Francisco to Singapore.[55]

***Graeber's Napping Experiment with NASA***

Graeber work included a NASA experiment and proved the positive effects of planned napping on long-haul airliner crews flying across the Pacific.[56]

In the experiment, four Boeing 747 three-member crews took 40-minute nap periods, one crew member at a time, during a quiet period of the flights. Three other crews acted as an experimental control, flying as usual without scheduled rest periods.

The crew members who were allowed to sleep for the brief nap periods tested out as significantly more alert during their landings than the control group. Graeber says that they now have dramatic unambiguous data showing that short rest periods improve the alertness of a crew when they begin the descent phase of their flight.

**Arguments Against Napping**

One argument against napping is that sudden arousal from a nap, as might occur due to an emergency will result in a serious performance reduction that could threaten safety. Laboratory tests show that shiftworker reaction time is slower during the first 5 to 10 minutes after awakening (the "sleep inertia period") and that deficits in problem solving can last up to 30 minutes.

Currently, it is against the rules to sleep in the cockpit on most U.S. airlines and certainly in most industrial jobs. Traditional managers say "sleep on your own time." But we can expect this to change due to results from NASA experiments and closely studied application of improvements.

However, napping is a normal activity in some 24-hour jobs, especially those jobs related to emergency response: firefighters, military security, and emergency medical personnel.[57] Japanese night workers already practice napping. Currently in most cases, American workers would lose their jobs if caught napping. Inevitably, we expect it to hit industry in this country within the decade.[58]

**A Napping Procedure**

According to the best thinking, the shiftwork nap should last from 10 to 40 minutes. The nap must be taken before the employee's sleep loss has accumulated to the point that it causes extended "sleep inertia" upon awaking. Sleep inertia is that initial slow reaction period upon awakening that can be dangerous on the job.

**Experimental Napping Procedure**

- Sleep 10-40 minutes.
- One person at a time.
- Sleep lightly.
- Watch for sleep inertia.
- Allow pre-sleep and post-sleep time.

Certainly, deep sleep isn't what we want during napping! Notice that NASA scheduled the aviator's experimental naps for about 40 minutes and for only one crew member at a time.[59]

**Improving Alertness with an Eating Procedure**

Dr. Judith Wurtman and other scientists at the very respectable MIT are developing breakthrough eating procedures that we should be trying on the front-line.[60]

### *A Finicky Brain!*

Wurtman has found that the brain is very selective about the amino acids that it accepts to facilitate learning, thinking, and staying alert. Your brain uses tyrosine to make the neurotransmitters that are apparently crucial to alertness, quick thinking, fast reactions, and memory. Alternately, it uses tryptophan to make the neurotransmitter that slows you down, makes you sleepy and mellow, and reduces your desire for control!

### *Racing at Amino Ups-or-Downs*

The funny part about this — it's an amino horse race to see which chemical the brain will accept after you begin to eat. After one amino chemtestant crosses the brain's finish line, the other is locked out. There is only one winner!

To increase alertness, Wurtman advises that we eat the protein portion (meats, no carbohydrates, high-protein cheeses) before taking a single bite of carbohydrate food (simple sugars and starches). Tyrosine will win the race and give you a mental boost.

**Experimental Alertness Eating Procedure**

- First, eat about 3 oz. protein.
- Then, eat carbohydrates.

### *Experimental Shiftworker's Alertness Eating Procedure*

In the many plants we visit, we suggest that shiftworkers try the alertness eating procedure to see how it affects their alertness. Before and during shifts, eat a portion (3 or 4 ounces) of meat, fish or poultry first. Then, eat or drink your carbohydrates (breads, pastas, sugars).[61] You can imagine their shock when we tell them to try eating the lean roast beef from their sandwich first before eating the roll! Low-fat jerky snacks or yogurt might ward off the sleepies throughout the shift. Sugars that are added to foods, i.e., jerkies and yogurts, may skew your efforts. In reverse, having a light carbohydrate snack before bedtime may help you sleep. A small bowl of cereal, cookies and milk, or a piece of toast. Be sure to track your results and look for improvements in your alertness, response time, or ability to sleep.

### *Reducing Jet Lag with Light Exposure*

NASA and Dr. Charles Czeisler were some of the first pioneers to experiment with installing intensive light in the workplace and studying its effect on alertness and performance. One "light treatment" experiment was recently performed at the Johnson and Kennedy space centers.

The experiment was carried out on the astronaut crew of Columbia (STS-35), that was scheduled to blast off during a night launch. This shuttle crew had already been divided into two crews: a 4-person red team and 4-person blue team, to perform shiftwork in space.

The red team had previously tried to invert their work/sleep schedules — to a graveyard mode — for two weeks but without any luck. In fact, they reported the usual complaints of the shiftworker, including: gastrointestinal disorder, difficult sleeping, sub-par working performance, and a pronounced "circadian dip" in alertness between 4 a.m. and 6 a.m.

For the experiment, the project officials exposed them to bright light in a "light room" at certain times during their "work shift" and also gave them a program on sleep hygiene, shiftwork and the internal clock.

At the end of the experiment, the researchers found that the red team had ceased to report the usual dip in alertness and had begun sleeping well again.[67]

## SUMMARY — PREDICTING FUTURE INCIDENTS

From safety statistics such as we have reviewed, we can pinpoint areas of future high incident probability — and these incidents will involve shiftworkers. Look at it this way:

If human error will account for most incidents in the future …
And if this error will tend to occur more after midnight ...
And if a high percentage of night workers will be working critical jobs...
And if this error will be committed more often by employees under two years ...
And if this error will occur more often during startups, shutdowns, maintenance, and shift changes ...
And if we don't change the system ...

Then, in a linear fashion, we can predict the details of some important (not too distant) future headlines for our own companies and others.

We can predict that a statistically high number of headline incidents will occur (or be set up!) after midnight due to human error and these incidents will involve one or more employees with under two years experience. Many of the incidents will involve set-ups such as poor fitness and health, overwork, and inappropriate meals.

These will have led to fatigue, jet lag, and slow response time — and will have caused a slip. Additionally in many industries, the slips committed during start-ups, shutdowns, maintenance and shift changes by newly-formed crews will make headlines!

## Predicting Future Incidents

- Headlines will report statistically high number of events occurring after midnight due to human error
- Employees with under two years experience will have more than their share
- Many accidents will be due to fatigue, slow response times
- Startups, shutdowns, maintenance activities and newly-formed crews will be involved.

These headlines will cost us great human and economic pain. As we have seen from the recent past, some of these incidents will be "world class" and will gravely impact organizations, lives, communities, and the environment.

Fortunately, these headlines aren't etched in steel and do not have to be written! By reducing shiftwork Set-Ups and Slips through better scheduling and performance improvement, we can rewrite our own headlines for the future.

## Rewriting Future Headlines

- Companies experience an 80% increase in employee fitness involvement!
- Lessened EAP and health costs due to new employee outlook on fitness vs. performance!
- Employees on their way to lowest level of loss-time accidents, and chemical spills and leaks
- Shiftworkers improve their performance and career outlook!

## CHAPTER THREE ENDNOTES

[1]Some consultants venture this human error figure to be 100 percent! W. G. Garrison has stated that except for natural disasters, human error is at the root of all losses if the investigator is willing to trace the sequence of events far enough. From: Major Fires and Explosions analyzed for a 30-year period by W. G. Garrison of M & M Protection Consultants, Chicago, Hydrocarbon Processing, September, 1988.

A 1988 unpublished "Safe Working Style" industrial study by Glenn McBride and Dow Chemical safety personnel on the *Incredible Chain of Events* concurs with this.

[2]From: Dr. David Dinges, biological psychologist, in Drowsy America written by Anastasia Toufexis, *Time*, December 17, 1990.

[3]See: Punch the clock, hit the hay — National Commission on Sleep Disorder Research finds evidence many industrial accidents happen due to lack of sleep. Merrill M. Mitler, *New York Times*, January 11, 1992.

Dr. Mitler writes that the U.S. is one of six industrialized countries that do not regulate the number of night shift hours most workers can perform. "In some transportation, health care and utility industries, the minimal U.S. regulations are full of loopholes and often ignored."

[4]See: Body-Wise: Safe Solutions for Night Work, by Susan Chollar, in the Competitive Edge Feature, *Psychology Today*, November 1989.

[5]"Fumbling slips" are those physical "motor errors" that are easy to observe: mis-sorting, mis-labeling, mis-operating — usually associated with touching/ grabbing/ flipping/ adjusting the wrong switch, lever, button, or other object!

[6]"Attention Slips" are those accidental slips of attention: mis-judging, switching and forgetting, unintentional micro-sleeping and macro-sleeping.

[7]See: A-B-Zzzzzzzz's of Sleep Cycles. Circadian study awakens hope for shift-workers, travelers, by Edmund L. Andrews, *New York Times* from The Houston Chronicle, April 30, 1990.

[8]Charles A. Czeisler is director of The Center for Circadian and Sleep Disorders Medicine at Boston's Brigham and Women's Hospital, Boston, Mass. An easy to read account can be found in: Night Shift Blues, Associated Press, May 8, 1990.

The statistical study included people driving front-end loaders, nuclear power plant operators, policemen, and chemical workers.

[9]From: Sleep cycles linked to major disasters, *Los Angeles Times*, 1988. The four incidents cited were: Three Mile Island, Chernobyl, Rancho Seco, and Oak Harbor. The research panel made recommendations that government, management, and labor become more aware of the "zones of vulnerability."

[10]From: Safety Gets Short Shrift on Long Night Shift, *Science News*, January 21, 1989.

[11]From: Soviets Okay $26 Billion in Resettlements, Associated Press, April 26, 1990.

[12]From: Beating the Clock — Was it an Accident Chernobyl Exploded at 1:23 in the Morning? *Wall Street Journal*, by Glynn Mapes, April 10, 1990.

[13]See: Punch the clock, Hit the Hay — Tired Workers Cause Millions of Accidents a Year by Merrill M. Mitler (San Diego), *New York Times*, Jan. 11, 1992. Dr. Mitler is director of Sleep Disorders at the Scripps Clinic.

[14]John K. Lauber often talks about the "ebb" of alertness and body temperature occurring between 4 a.m. and 6 a.m. in the morning. John Lauber is a Member of the National Transportation Board, formerly with NASA, and investigates major transportation disasters. He has been supportive of industrial programs on human error.

[15]See: Drowsy Americans by Anastasia Toufexis, *Time*, December 17, 1990.

[16]From: Does Shiftwork Shift Risks? by Michael K. Coplen, Safety and Health, June, 1988. In this article researcher and consultant Dr. Coplen states that "80 percent of shiftworkers experience personal, social and medical problems" related to shiftwork.

[17]From: Human Factors in Aviation, Aircrew Fatigue and Circadian Rhythmicity by R. Curtis Graeber, January 1, 1988.

[18]According to Lidia Dotto, a sudden eight-hour shift produces about as much circadian disturbance as flying from the middle of North America to Europe. From: *Losing Sleep — How Your Sleeping Habits Affect Your Life* by Lidia Dotto, Morrow Publishers, 1990.

[19]Peggy tells new-hires that "jet lag feels like a case of the flu." So far, the experienced shiftworkers present at the events haven't challenged her on this!

[20]Reported in *Business Week*, February 18, 1991.

[21]No standard method of monitoring and recording "body clock time" of our 24-hour employees has been developed.

Pilots frequently record their "body clock time" on incident reports. This helps us interpret their fatigue and jet lag reports in the "Aviation Safety Reporting System" maintained at NASA Ames Research Center, Moffett Field, California. They simply estimate body time based their awaking at a theoretical 6 a.m. body time. We need something like this for our other 24-hour jobs!

[22]From: *How to Make Shiftwork Safe and Productive* by Timothy H. Monk, Ph.D., published by American Society of Safety Engineers, 1800 East Oakton Street, Des Plaines, IL, 1988.

[23]*From: personal communications with James C. Miller, Ph.D., 1992.* Dr. Miller has written extensively on vigilance, human performance and safety. Dr. Miller consults privately from the San Diego, California area. He has conducted U.S. Dept. of Transportation and industrial investigations of driver fatigue, alertness and performance in truck and bus operations and co-authored a literature review of human vigilance performance.

At the USAF School of Aerospace Medicine, he investigated aspects of aircrew and ground personnel work load, performance and fatigue, and work station and work shift design. Dr. Miller was Chief of Human Factors for USAF development flight test at Edwards AFB 1987 to 1989.

[24]One report of this was found in the Associated Press: Utilities average one fine a week at nuclear reactors, November 28, 1991. Sleeping at the controls was not considered an “immediate threat to residents around the plants.” This intentional human error was classified a little more critically by residents living near the operation!

[25]From: Upstate New York’s Wreck of the Champlain by Derek R. Grout, *Skin Diver*, June, 1989.

[26]From: F-117A Crash Reports Cite Pilot Fatigue, Disorientation by William B. Scott, *Aviation Week & Space Technology*, May 15, 1989.

[27]We owe this excellent picture to an interesting 1992 talk with Bill Barringer of Conrail. Bill’s *rail craft scene* struck Glenn as having definite analogies with the NASA Apollo flights. Crews on longer missions absolutely must depend upon their own fitness, alertness, and training. NASA learned from the first manned trip around the moon that a crew with high fatigue, terminal jet lag, and on a carnival director’s schedule could not function well enough to complete the mission competently. Peggy tells classroom participants how we peeled them out of the returning space capsules in a “Jello condition."

[28]The OTA report stated that because of this, improved training is crucial—especially for the 1.7 million local fire and emergency personnel, who are first on the scene of spills. From: *U.S. News & World Report*, July 21, 1986. Transportation excitement mentioned on the same day of this government release was the tank car releasing white phosphorous vapor over Miamisburg, resulting in 25,000 evacuating their Ohio residences.

[29]See: Drowsy Americans, Anastasia Toufexis, *Time*, December 17, 1990.

[30]See: *Wide Awake at 3:00 A.M. — by Choice or by Chance?* by Richard M. Coleman, W. H. Freeman and Company, New York, 1986.

[31]See: Drowsy Americans, Anastasia Toufexis, *Time*, December 17, 1990.

[32]From: *The Sleep Better, Live Better Guide*, by the Better Sleep Council, P.O. Box 13, Washington, D.C., 20044.

[33]From: Report adopted by the National Transportation Safety Board, February 5, 1990.

[34]From: Why drivers fall asleep, and how to avoid becoming a traffic statistic by Jane E. Brody, *New York Times*, July 5, 1990.

[35]Most label micro-sleep as that unintentional sleep lasting from a few seconds to three minutes. Micro-sleep is the body's strategy to catch up on sleep, in spite of the shiftworker's desire to stay awake continuously.

[36]From: a 1992 Associated Press report (McKinney, Texas).

[37]From: *Christian Science Monitor*, October 4, 1989. According to the article, pilot error was involved in 63 percent of all airplane accidents involving large carriers and commercial flights in 1985, the latest year of statistics from the National Transportation Safety Board at that time. We think this figure included all cockpit crew members.

[38]Much has been done by the National Transportation Safety Board and by NASA research.

Our safety training programs, *Safe Working Style* and *Unsafe Act Intervention*, apply much of what has been discovered in government research to industrial settings. Both programs were developed through Dow Chemical and they have been the basis of their corporate safety culture since their original release.

[39]See: *Wide Awake at 3:00 A.M. — by Choice or by Chance?* by Richard M. Coleman, W. H. Freeman and Company, New York, 1986.

[40]From: *America in the Global Nineties* by Austin H. Kiplinger and Knight A. Kiplinger, Kiplinger, Washington Editors, 1989.

[41]In October 1991, Congress approved $2 billion for the space station, which NASA plans to have in orbit and operated by four resident astronauts before the year 2000.

[42]From: Conversations with NASA, 1992.

[43] Quoted from: *Report of the Presidential Commission on the Space Shuttle Challenger Accident, Volume II* (often called "The Rogers Report").

[44]Dr. Richard Coleman tells us that a sleepy night technician had been working a number of consecutive 12-hour shifts. From: *Sleep Alert*, a PBS television broadcast, produced by Dr. James B. Maas, Psychology Film Unit, Cornell University, underwritten by Abbott Laboratories, March 22, 1990.

[45]From: Rogers Report Cites Dangers of Excessive Overtime by Craig Covault, *Aviation Week and Space Technology*, August 21, 1991.

[46]In summarizing, the Rogers Report ties in *time pressure and human error:* "Time pressure, particularly that caused by launch scrubs and rapid turnarounds, increases the potential for sleep loss and judgment errors. This could be minimized by preventing launch support personnel, particularly managers, from combining launch support duty with office work on the same day. The willingness of NASA employees in general to work excessive hours, while admirable, raises serious questions when it jeopardizes job performance, particularly when critical management decisions are at stake."

[47]From: A conversation with Keith Fox, New Orleans American Petroleum Institute on Training Development, Fall Conference, 1991.

[48]See: Drowsy Americans, Anastasia Toufexis, *Time*, December 17, 1990

[49]Four years ago, health officials in New York limited the on-duty hours of emergency room physicians after a young woman died soon after being admitted to a hospital. Two doctors who treated her had worked 18 hours. A grand jury recommended the regulations which are expected to cost the state $3.1 billion over the next 10 years. From: Shiftworkers Face Thorny Problems by Ann Levin, Associated Press, June 10, 1992.

[50]Could it be that "police abuse" leads to police abuse? It doesn't seem too far-fetched to some of the folks we've talked with!

[51]William C. Cunningham is President of Hallcrest Systems, a law enforcement research firm. He has written a 1990 U.S. Department of Justice study. From: Working in the Moonlight — 85% of HPD officers work after-hours jobs, by Karen Roebuck, *Houston Post*, September 14, 1991.

[52] (or "cooping")

[53]See: Why drivers fall asleep, and how to avoid becoming a traffic statistic. Jane E. Brody, *The New York Times*, July 5, 1990.

[54]See: *Losing Sleep, How Your Sleeping Habits Affect Your Life* By Lidia Dotto, Morrow, NY, 1990.

[55]See: Crew Fatigue Monitoring by David F. Dinges and R. Curtis Graeber, *Flight Safety Digest*, October, 1989.

[56]From: Study Finds In-Flight Rest Periods Are Beneficial to Pilots on Long-Haul Flights by Breck W. Henderson, *Aviation Week & Space Technology*, July 9, 1990.

[57]Jamal found that firefighters were not seriously affected by rotating shifts like the other shiftworkers in his empirical study. *He attributes this to the traditional napping allowed on the job.* From: Shiftwork Creates Unusual Problems, by Muhammad Jamal, *Personnel*, May, 1989.

[58]Organizations experimenting in this area may contact and update McBride Publications at the address in the back of this manual.

[59]Dr. Timothy Monk tells supervisors that sleep is more refreshing when there are no interruptions and napping should be used only as an emergency "fill-up" process. From: *How to Make Shiftwork Safe and Productive* by Timothy H. Monk, Ph.D., published by American Society of Safety Engineers, 1800 East Oakton Street, Des Plaines, IL, 1988.

[60]From: *Managing Your Mind and Mood Through* Food by Dr. Judith Wurtman, Harper and Row Publishers, 1992. (Original edition was released in 1986 by Rawson Associates.)

[61]If this works for a given individual, then the opposite should work: eating carbohydrates first to wind down and relax!

[62]A easy-to-read report is found in Coping with your Body Clock by Peter J. Sheridan, *Occupational Hazards,* November 1991. (Sheridan also mentions that Dr. Czeisler has designed a two-day course, *Work Schedules and the Biological Clock* at Harvard University of Public Health. Phone 617-432-1171.)

Timothy Monk writes that the hormone "melatonin" was studied in the eighties. This hormone makes one sleepy when it is circulating in the blood, therefore it appears to influence the internal clock. Peak levels of melatonin in the blood are found at night and the hormone can be wiped out by exposing the individual to bright light. From: *How to Make Shiftwork Safe and Productive* by Timothy H. Monk, Ph.D., published by American Society of Safety Engineers, 1800 East Oakton Street, Des Plaines, IL, 1988.

Researchers at the Oregon Health Sciences University plan a $4 million project to study melatonin's role in controlling sleep and wakefulness. They believe the hormone's level declines as individuals age which disrupts the body's internal clock. Robert L. Sack, principle researcher, says that people in their thirties are affected and don't have the same deep sleep of younger people. It is estimated that half of the 29 million people over 65 have trouble sleeping. From: Associated Press release, Thursday, July 9, 1992.

# CHAPTER FOUR

## Scheduling Shiftworkers

### OBJECTIVES OF CHAPTER FOUR

By the time you have finished this chapter, you will be able to debate and discuss different rotating schedules and their impact on employee safety, performance and lifestyle.

### A MULTITUDE OF SCHEDULES

This chapter will give you a primer on shift scheduling. Researchers find that the weekly pattern of shift rotation has become a norm in American continuous process industries.[1] In spite of this, Dr. Richard Coleman, a California sleep expert, believes that most American shiftwork schedules have become outdated. He has stated that the results of these outdated schedules are low productivity, high overtime, reduced alertness and safety, and poor teamwork. We agree with this.

## Using the Worst Schedule

The Reverse Rotating 8-hour Schedule (Evenings — Days — Graveyards) was the most common in America just a few years ago.[2] It has also been called the "southern swing" schedule and is considered the poorest rotating schedule (of 42 hours per week).

In addition to a fifteen-day stretch with only one day off, the schedule fights the human's internal clock. We believe that many reactions against shiftwork are due to this one schedule![3]

**8-Hour Reverse Rotating**

| Sun | Mon | Tues | Wed | Thurs | Fri | Sat |
|---|---|---|---|---|---|---|
| | | E | E | E | E | E |
| E | E | | D | D | D | D |
| D | D | D | | | G | G |
| G | G | G | G | G | | |

D = Days (typically 7 a.m. to 3 p.m.)
G = Graveyards (typically 11 p.m. to 7 a.m.)
E = Evenings (typically 3 p.m. to 11 p.m.)

### *Navy Nuclear Sub-Standard Schedule*

A 6-hour schedule on board U.S. Navy nuclear submarines could possibly be vying for overall worst schedule award! We understand that there is a great turnover of personnel on this schedule. Of course, there may other reasons for the turnover.

6 hrs. work
6 hrs. sleep
6 hrs. training, study and free time

| 18-Hr Day #1 | | → | 18-Hr Day #2 | | |
|---|---|---|---|---|---|
| Work<br>6 | Sleep<br>6 | Train<br>6 | Work<br>6 | Sleep<br>6 | Train<br>6 |

### *Using the Right Schedule!*

Preston Richey, a scheduling consultant, tells us that using the right schedule for a certain group of employees is critical. It's a problem that the typical company underestimates.[4]

Richey's previous company, Salt Lake Minerals and Chemical Corporation, changed to a 12-hour schedule. He says the one thing that they could agree on in the beginning was that they hated their original schedule.

### *Favoring 12-Hour Features*

Although we don't promote one particular schedule — certainly no 8-hour schedule, we do favor the 12-hour schedule's features for most chemical plants and refineries. Although it may prove tougher on supervisors, employees experiencing it develop a much more positive viewpoint toward 24-hour work ... and their overall jobs. This reception to the 12-hour schedule is repeated hundreds of time each year across America as plant sites try it, often due to employee suggestion.

## 12-HOUR SCHEDULES

Richard Coleman says that the 12-hour schedule came to America in 1955.[5] By the 1970s, rotating schedule workers in chemical plants and refineries were hearing about it and wondering if it would get them away from six- and seven-day schedules. Locations switching to 12s during the 70s and 80s included companies like Borg-Warner, Ciba-Geigy, Exxon, Eli Lilly, Du Pont (Richmond, Virginia), Raritan River Steel (New York), Bridgestone/Firestone (Warren County, Tennessee), AMAX Coal (Gillette, Wyoming), Dow Chemical - Oyster Creek Division (Freeport, Texas) and Great Salt Lake Minerals and Chemical (Ogden, Utah).

### Most Popular 12-Hour Schedules

Even though there are hundreds of 12-hour schedules being used today including combination 12s and 8s, two forms remain most popular. One has a long (6 or 7)-day break and the other form limits work nights to three in a row and all breaks to a minimum of two days.[6] The second type is sometimes called the "work a little, rest a little, every other weekend off" schedule.

Employees at the Exxon refinery at Baton Rouge call this the "EOWEO Schedule" (Every Other Weekend Off) and have worked it steadily since February, 1975.[7] Employees within Exxon refer to this schedule as the "Exxon Schedule;" employees with Fina consider it to be the "Fina Schedule." In any case, this schedule rotates six times in one month and is extremely overtime intolerant. Typically, it does not provide a good relief system. Many companies have made variations to this original version.

**12-Hour Alternating EOWEO**

| Sun | Mon | Tue | Wed | Thu | Fri | Sat |
|---|---|---|---|---|---|---|
| D | | | N | N | | |
| | D | D | | | N | N |
| N | | | D | D | | |
| | N | N | | | D | D |

The EOWEO has several strong features:

- It repeats consistently on a four-week cycle.
- All shifts last exactly 12 hours.
- Every other weekend features three days off.
- A maximum of three work nights in a row.
- This schedule is a fast rotating schedule — 6 rotations in a 28-day period.
- This schedule is overtime intolerant.
- Slowing the rotation to twice in one month can make this schedule more circadian friendly.

### *12-Hour/4-Day/4-Crew Schedule*

The 12-Hour/4-Day schedule is popular in many refineries. Although it requires some payroll policy acrobatics, it provides a consistent, predictable schedule for everyone involved. Some circadian specialists would worry about the four nights in a row.

Individuals working this schedule say that they experience the greatest alertness dip on the second night. Predicting a social life can be difficult on this schedule due to the length of the 8-week cycle.

12-Hour 4-Day, 4-Crew

| Sun | Mon | Tue | Wed | Thu | Fri | Sat |
|---|---|---|---|---|---|---|
| D | D | D | D | | | |
| N | N | N | N | | | |
| | | D | D | D | D | |
| | | N | N | N | N | |
| | | | | D | D | D |
| D | | | | N | N | N |
| N | | | | | | D |
| D | D | D | | | | N |
| N | N | N | | | | |
| | D | D | D | D | | |
| | N | N | N | N | | |
| | | | D | D | D | D |
| | | | N | N | N | N |
| | | | | | D | D |
| D | D | | | | N | N |
| N | N | | | | | |

**Employee Involvement is Where it Begins!**

Twelve-hour schedules are usually promoted and planned by the local plant shiftworkers. The typical conversion to a 12-hour system goes through an employee involvement and voting process. This "bottom-up" approach often takes more than a year to "negotiate" with management and usually involves a 12-hour schedule committee.[8] The schedule then undergoes a trial period where employees try it and supervisors monitor the change.

**Employee Consensus Continues the Schedule**

After the trial period of 3 months to a year, employees once again vote on whether they really like the 12-hour schedule. Invariably they do. Also ... invariably most supervisors (who continue to work days) do not — or at least do not appreciate it for the first year or two![9]

**New Schedule Anxiety**

Setting up a trial run of the 12-hour schedule is a real anxiety producer for all involved. During the pre-schedule and initial implementation phases, initial concerns include: increased fatigue of workers, lower productivity, tougher communications, relief crises, extra costs, and time for training. After a couple of years, most fears are allayed or put on the back burner.

### Initial 12-Hour Schedule Concerns

- Increased Farigue
- Lower Productivity
- Tougher Communications
- Relief Crises
- Extra Costs
- Time for Training

**12-Hour Fatigue**

The longer workday becomes a primary concern of supervision during initial discussion. It is presumed that much greater fatigue will occur during the latter part of the 12-hour day. Considering just one day independent of an overall work schedule, this makes sense. And literature is available to support this idea. Many past 8-hour shifters attribute their fears of fatigue based on the concept of 8-hours regular shiftwork plus 4-hours overtime, without assessing the entire schedule.

Studies of radar operator fatigue during WWII as well as aviation and ground transportation studies over the last 20 years show that accidents increase markedly during the second half of the shift. It is tempting to apply these statistics to the 12-hour schedule.

However, 12-hour safety success stories abound. The AMAX Coal Company found that its safest period was the last four hours of the 12-hour shift.[10] Additionally, one mine at the operation was voted the safest open-pit location in the country after switching to their new 12-hour schedule.[11]

**Our Position on 12-Hour Fatigue**

In spite of management's concerns about greater fatigue on 12-hour shifts, we feel that accidents occur more frequently after midshift whether the employee is on 8s, 10s or 12s. And we feel that employees on each of these schedules can be equally subject to a Danger Zone dip.

Furthermore, we realize that we have a human case of apples and oranges. An experienced shiftworker will quickly point out the following: An employee working four 12s following four days at home can't be compared directly to an employee working seven 8s after only 1-1/2 days at home! The 12-hour employee may very well feel better and more alert as each shift change nears. In fact, the average 8-hour employee will be experiencing a good case of jet lag at work on several of his/her seven days, nights, and evenings.

**Have You Heard of the 8/12-Hour Schedule?**

Also, the "greater fatigue" thinking runs against a management reality that we find in many plants. Working purported "8-hour schedules," many energetic employees wind up writing an amazing number of "12s" on their time cards. "Call-outs" may occur so frequently that they result in a de facto combination 8/12 hour schedule existing on a 5 to 7-day rotation. At such sites, only 16-hour days are considered out of the ordinary and beyond the call of duty.

### The 8/12 Hour Schedule!

Unanticipated 4 Hrs. OT on
top of regular 8 Hrs. Due To:

- Stayovers
- Callouts
- Startups
- Shutdowns
- Turnarounds

Even the infrequent rotation breaks may be worked during start-ups, shutdowns, turn-arounds and high production runs. So, at these sites, a maverick sort of 12-hour schedule already exists — and with short, insufficient internal clock rest breaks between shift rotations! Many plants both union and non-union are working unsafe hours such as these:

**8/12 Hour Schedule**

| Sun | Mon | Tue | Wed | Thu | Fri | Sat |
|---|---|---|---|---|---|---|
| | 8-R<br>4-OT | 8-R<br>4-OT | 8-R | 8-R<br>4-OT | 8-R<br>4-OT | 8-R<br>4-OT |
| 8-R | 8<br>callout | 8<br>callout | 8-R | 8-R<br>4-OT | 8-R<br>4-OT | 8-R<br>4-OT |
| 8-R<br>4-OT | 8-R<br>4-OT | 8-R<br>4-OT | 8<br>callout | 8<br>callout | 8-R<br>4-OT | 8-R |
| 8-R<br>4-OT | 8-R<br>4-OT | 8-R<br>4-OT | 8-R<br>4-OT | 8-R<br>4-OT | | 8<br>callout |

(R = Regular) Total Regular Hours on 8 hour shift: 168 hrs
(Callout on dayoff) Total Callout Hours: 40 hrs
(OT = Overtime) Total Regular OT: 64 hrs
Fatigue and Stress Levels: Extreme

### 12 Hours ... A Hot Subject for Some

Some industries have a real reason to take heed of the 12-hour fatigue issue. Foundry workers and steel workers may suffer overexertion due to the heavier work and extreme heat. Due to this environment, many do not consider 8-hour-plus workdays to be safe. Many alternate from 8s to 12s seasonally.

### Concerns about Lower Productivity

Lower productivity goes along with the fatigue hypothesis. It is often presumed that a 12-hour worker will dilute job performance over the longer work period. Once again, our opposing position is that the greater morale, reduced jet lag, fewer shift changes, more restful time off, and less commuting lead to greater productivity by the average employee.

Certainly, if we can eliminate burnout and the Double H Pattern, we will improve productivity! AMAX Mining reported a 12 to 14 percent productivity improvement after switching to 12s.[12]

### Concerns about Tougher Communications

Communications definitely become tougher to manage when employees regularly disappear from three to seven days at a time! CYPRESS - AMAX's Ron Spangler says that it is very difficult to talk to all workers quickly. He reports that it can take a week to get around to everyone.[13] By arranging meetings pre-shift for each crew, you can cover the plant within a week. We'll bring this subject up again in a minute.

### Concerns about Relief Crises

The 12-hour system requires a special focus on how employees are to be relieved during vacations and emergency illnesses. The 4 a.m. call-out is no longer an option for schedulers! This can cause great anxiety to the strongest of our shift management.

Under "The Fifth Crew," we will discuss a powerful solution to relieving on 12s.

### Concerns about Extra Costs

With the many rules and restrictions on compensation across the country, it would be easy for the 12-hour schedule to increase payroll, overtime lunch, and other costs. These must be handled on an individual basis, depending upon whether the site is unionized, depending upon the state, and depending upon the size of the organization.

Many companies make it a requirement that the 12-hour system will not cause any significant cost increases. Others go along with some increases in order to reap the benefits of the more palatable schedule.

### Concerns about Training

One area that has really suffered for many companies implementing the 12-hour schedule is simply finding a good time to schedule high quality training and safety meetings. Supervisors quickly find that to hold a training or communications session, their operators must be asked to stay past 12 hours, come in early, or ... come in on a day off. None of these requests are accepted well by the employees to say the least! In fact, attempts at scheduling such sessions will quickly elicit a disheartened protest.

## THE FIFTH CREW AS THE TRAINING SOLUTION

This quandary has been totally overcome by organizations that have implemented a "fifth crew" and we strongly recommend the format. With a fifth crew, all personnel are required to cycle through a training and relief schedule every five weeks. Glenn designed (and experienced) the fifth crew format on his 12-hour schedule work in the 70s at Dow Chemical and it worked out quite well. The training - communication - relief - maintenance weeks were held back-to-back every 10 weeks and were scheduled as two weeks of 8-hour days.

### Unwinding the Clock!

The fifth crew worked out better than anyone had imagined during our months of planning. One unexpected payoff was a continuous period of straight day work ... four weeks of 12s running the plant and then two weeks of 8-hour days undergoing training, performing running maintenance, and acting as reliefs for vacations, etc.

Additionally, the more experienced rotating Technicians used the extra 8-hour days to catch up on writing training manual revisions, performing inspections, and doing other special tasks — depending upon the make-up of the crew. Although, initially the extra crew seemed an added expense to our personnel manager, no one was ever caught with slack time during those fifth weeks. The fifth crew performed much of the maintenance and project load that previously had been assigned to a "repair crew." It also eliminated any extra relief personnel. Some plants who have downsized have optimized fifth crew for cross-training operations as maintenance personnel and vice versa.

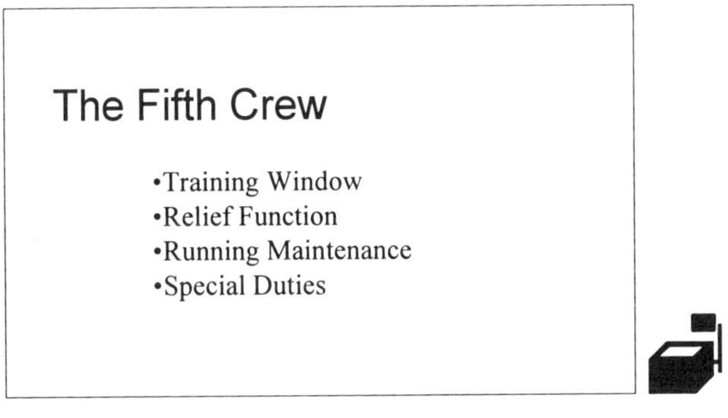

The following three schedules are examples of 12-Hour Schedules with Fifth Crews that have proven quite satisfactory. The 6-Day Break Schedule has the advantage of a 6-day break during a stretch of days. This encourages vacations to be taken during day work rather than on night work. [14] The ten-week cycle includes a four-week stretch of night work. However, this is followed by six weeks of day work to recoup the internal clock!

## 12-Hour Fifth Crew 6-Day Break

| Sun | Mon | Tue | Wed | Thu | Fri | Sat |
|---|---|---|---|---|---|---|
| | 8 | 8 | 8 | 8 | 8 | |
| | 8 | 8 | 8 | 8 | 8 | |
| | D | D | D | | | |
| | D | D | D | D | | |
| | | | | D | D | D |
| D | | | | | D | D |
| D | | | N | N | N | |
| | N | N | | | | N |
| N | N | N | | | | N |
| N | | | N | N | N | |

8 = Eight-hour day (Fifth Crew work for training, relief)
D = Twelve-hour day
N = Twelve-hour night

Our next schedule example, the Split Nights Schedule, breaks up the night stretch with two weeks of 8-hour day work:

## Fifth Crew/Split Nights Schedule

| Sun | Mon | Tue | Wed | Thu | Fri | Sat |
|---|---|---|---|---|---|---|
| | 8 | 8 | 8 | 8 | 8 | |
| | 8 | 8 | 8 | 8 | 8 | |
| | N | N | | | | N |
| N | N | | | | | N |
| N | | | D | D | D | |
| | | D | D | D | D | |
| | D | D | | | | D |
| D | D | | | | | D |
| D | | | N | N | N | |
| | | N | N | N | N | |

8 = Eight-hour day (Fifth Crew work for training, relief)
D = Twelve-hour day
N = Twelve-hour night

The fifth crew 3-day schedule features three-day on/off sequences and completes a total cycle every 15 weeks. It maintains the consistency of the four on/four off schedule. This schedule is tough on spouses and social lives.

## Fifth Crew/3-Day Schedule

| Sun | Mon | Tue | Wed | Thu | Fri | Sat |
|---|---|---|---|---|---|---|
| | 8 | 8 | 8 | 8 | 8 | |
| | 8 | 8 | 8 | 8 | 8 | |
| | 8 | 8 | 8 | 8 | 8 | |
| | D | D | D | | | |
| N | N | N | | | | D |
| D | D | | | | N | N |
| N | | | | D | D | D |
| | | | N | N | N | |
| | | D | D | D | | |
| | N | N | N | | | |
| D | D | D | | | | N |
| N | N | | | | D | D |
| D | | | | N | N | N |
| | | | D | D | D | |
| | | N | N | N | | |

8 = Eight-hour day (Fifth Crew work for training, relief)
D = Twelve-hour day
N = Twelve-hour night

We prefer fifth crew scheduling (and you may find your department coming up with more variations.) All in all, the fifth crew 12-hour schedule meets the increased demands of the nineties: reduction of jet lag, reduction of travel and commuting load on the environment, and ... a day window for training, communication, and employee development.

## 12-Hour Schedule Advantages

In spite of the usual initial concerns, all 12-hour systems are usually evaluated highly over the long term. Advantages include:

### 12-Hour Schedule Typical Advantages

- Improved morale
- Reduced jet lag and fatigue
- Increased productivity
- Fewer evening or graveyard shifts
- Blocks of time-off better
- Double weekends off
- Improved family relations
- Fewer and better shift changes
- Reduced commuting, less trips to/from work
- Reduced absenteeism

## Improved Morale

The greatest effect of switching to 12-hour shifts is usually an improvement in morale for employees and supervisors. Employees experience a refreshing change with shorter work sets, and spouses notice the employee is home more often and for longer blocks. Some employees go through a transitional phase finding hobbies and off-hour activities to fill their time. A few complain that they spend too much money on their time off!

It's important to remember that the schedule was originally created only to improve the working conditions, energy level, and job satisfaction of 24-hour workers.

## Reduced Jet Lag and Fatigue

Jet lag and long-term fatigue are often reported as reduced. As long as employees work and sleep while on 12-hour shifts and save off-hour activities for days off, even long-term jet lag improves with the newer scheduling. Older habits from 8-hour shifts must be addressed and possibly retrained if the change is to be successful. You will hear complaints from employees that "there is no time to do anything else," which is basically true, except for simple personal duties and perhaps a load of laundry or two. Employees must look for support systems to make the 12-hour shiftwork, i.e., cleaning help, childcare, cooking, automobile maintenance, lawncare, second job responsibilities. There is no "slack" time on the 12-hour schedule and work days must be optimized.

A good exercise Peggy uses in training workshops is termed "Get a Life." It's a behavioral simulation and nearly a Level 4 evaluation (for you trainers) where attendees apply the skills learned in the workshop to an employee profile. The profile includes years of employment with a ghost company, marital status and problems, children ... and their problems, health issues, and work issues. The profiled employee is also a shiftworker, perhaps on 8s, 10s, 12s or a combination, working in a critical 24-hour environment. It is up to each team to plan a Performance Tracking program for their "Life" that encompasses each of the F.A.S.T. Tracking performance areas, provides measurable results, and is locally achievable.

Some of the materials included with the exercise may be local travel brochures within a short distance of the "Life's" home, self-help clippings from local newspapers, computer bulletin board or Internet addresses, and other social activities.

In several plants, we have even videotaped the role play and left the videos with the employees. It turns into a rowdy time, but obvious learning has occurred and the attendees are applying their newly learned skills at improving vigilance, fitness and health, sleep, quality of time off, scheduling assessment and intervention abilities all at the same time. This is an excellent exercise.

Many employees will identify with their random "Lives" so well that they even name them. Managers who have doomed their plants as hopeless and helpless, employees as burned-out, not motivated, will usually marvel at the new-found strength and enthusiasm in their front-line operational employees as they learn to take control and make changes.

**Increases Productivity**

Many companies report greater productivity after switching to the 12-hour work day. Certainly, if productivity and morale go hand-in-hand, then this is true in many cases. This is a controversial item among management.

A common report we track is that with less shift changes in a 24-hour period, there are less hands on the controls! There is less variation in computer set data points. This in itself increases product quality and increases product output.

Also, employees feel it is easier to come in to work two to three days in a row. They are not dreading the fifth, sixth, seventh 8-hour shift day.

### Fewer Evening and Graveyard Shifts

Many employees see rotating shiftwork as being only one-third day work. The other two shifts appear as off-track or night shifts to them. When working 12s, the number of days increases by 50 percent and the number of nights decreases by 50 percent.

### Blocks of time off Better

An additional 78 days off per year is gained on the twelve-hour shift. Greater advantage can be taken of the larger off-time blocks. Often, taking 36 hours of vacation time will reap 10-plus days of calendar vacation. In Peggy's F.A.S.T. Tracking sessions, special attention is given to scheduling time off, sharing calendars with close friends and relatives and even scheduling sleep time so that friends know when to call.

There is a much better chance that any given holiday will occur within a given block of time off. This works well for employees who want to visit extended families living out of state. The forty to fifty percent of holidays falling within major blocks of time off delight this group.

### Doubled Weekends Off

The 12-hour worker averages every other weekend off as opposed to the every fourth weekend off of the 8-hour employee. Additionally, these weekends, like the holidays, may fit within two- to four-day blocks of time off.

### Improved Family Relations

Family relations are potentially improved. The evening meal can be scheduled with the family on all calendar days. One spouse remarked: "He can finally eat dinner with us 356 days a year!" Also, the family is not awakened by the shiftworker coming and going during the night!

The employee, with larger blocks of work and free time should experience an improved family life. However, it becomes a matter of keeping it under control. These blocks of time can be put to great use or can cause a family disintegration if abused.

### Fewer Shift Changes

Again, fewer shift changes reduce the exposure of lost communications (see "jet lag meets fatigue"). Plus, fewer shift changes mean less lost time "winding up" before shift change or getting into gear after shift change.

Since there are only two shifts working per day, theoretically communications load between the shifts will be reduced by 33 percent. Fewer shift changes also mean less trips to work and less exposure to other hazards on the road.

### Increased Shift Change Alertness

An interesting discussion centers on the 12-hour schedule's hours of shift change. Moving the hours of relief away from the two Danger Zones may improve the quality of shift change. Typically, hours of shift change are 6 a.m./6 p.m. or 7 a.m./7 p.m. Some mining operations are running a successful 12 p.m./12 a.m. shift change. Supervisors and training personnel enjoy the 12/12 start/stop time since they can communicate with two shifts on the same day without holding over! If a 12/12 start and stop time is the consensus at your plant, be sure to discuss the family impact. Unreal expectations of the shiftworker may bring about unfavorable results. Be sure to consider all areas of the 24-hour lifestyle. Sleep must be given priority. Breaking up the sleep period for quality family time should be avoided at all costs.

### Increases Shift Change Responsibility

One advantage occurs on days that the employee must relieve the person that he/she has previously relieved. "Chaining" this responsibility motivates the employee to leave the work area in the best of shape at each shift change.

Employees working eights carry around numerous shift change tales about individuals and crews they have dreaded to relieve in the past. These tales usually target employees that they have relieved regularly, yet who seldom or never were required to relieve them.

Another interesting comment concerning shift change responsibility is that housekeeping can even improve. Employees feel that are not being rushed and that they have the few extra minutes to sweep or water wash an area.

**Reduces Commuting**

Commuting drops from the 5-1/4 round trips per week to 3-1/2 trips per week. Therefore, commuting expenses are reduced significantly. And, moving shift change hours away from one or both Danger Zones may very well improve driving ability.

**Improved Relief System**

The temptation to ask relief employees to stay over until 4 a.m. or the flip side, to have them come in at 4 a.m. is theoretically eliminated.

This comes at a price, of course. An extra someone will have to appear at 7 a.m. or 7 p.m. Who will it be? Probably someone on standby, on the "fifth crew," or a day worker. Luckily, relief ethics quickly become very tight on twelves. Sick personnel working nights should call in very early.[15]

**Reduced Absenteeism**

As Raritan River Steel Company's John Browning says, "There's a 30 percent greater chance of a person getting sick on their own time." He stated to Industry Week that 8-hour workers will sometimes call in sick to take care of personal business because they have so few days off.[16]

**Reduced Pressure for Day Jobs**

In most cases, the pressure to transfer to day jobs is completely eliminated. For once, the shiftwork job becomes the preferred job to that of a day job. Scheduling jealousy may become an issue with other departments, i.e., engineering, management and maintenance. Perhaps you might want to look at improving scheduling for these departments after you're settled in. Suggested schedules might be 10s, 12s or even the 9-80.

**Running a Tight Ship**

For all of it's reported advantages, the 12-hour schedule has one Achilles Heel. It requires running a very tight ship by both managers and employees. All must learn to work along very disciplined lines: surprise time off must be minimized; commuting time must be handled efficiently; and vacation schedules must be handled with a greater discipline and lead time.

Employees quickly find the need for the discipline. A few extra hours at work, an hour of shopping on the way home, a movie on a work day... and one is instantly out of synch and fighting to get back into the 12-hour work/sleep rhythm.

**12-Hour Schedule Pay Issues**

Pay issues take a good deal of work when planning for conversion to the 12-hour schedule. The following issues are examples.

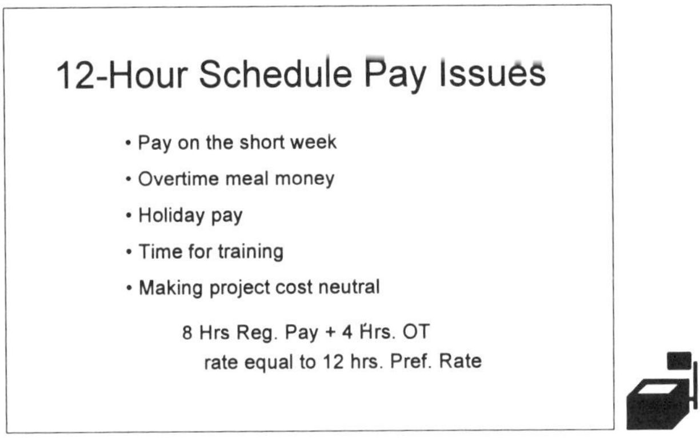


**Working the Short Week**

If the employee works on his/her day off on a short week ... will he/she be paid straight or overtime before the first 40 hours are worked?

Many companies reduce the hourly pay by some percentage in order to pay about the same overall.

**Handling Overtime Meal Money**

Will the employee receive an overtime meal allowance if he/she works 13 hours on a twelve-hour day? (In many cases, this is the rule.)

Also, if the employee works on his/her day off, when will overtime meal allowance be paid? (In many cases, it is paid only after working twelve hours.)

### Holiday Pay

How many hours holiday pay does the employee get when working a 12-hour day? A typical scheme is: 8 hours at 1-1/2 time plus 4 hours at 2-1/2 time plus 8 hours of standard time for a total of 30 hours straight time for the entire day. Of course, this is from companies on the high side of compensation. However, employees off that day or those cycling through eight-hour weeks would not receive anything near this.

How many hours holiday pay does the employee receive when off on a holiday?

### Time for Training

Pay for training time can develop into an issue. Will individuals coming in for training on their day off in a short week be paid nothing, straight or overtime pay for the training time?

### Cost Neutral

If the site must conform to strict state or union guidelines, you will run into some base pay factoring issues so that the eight hours plus 4 hours of overtime each day equals 12 hours of regular pay at the old rate. This new rate will be called the preferred rate. Other issues to be considered include shift differential and factoring double and triple pay issues.

## Final Comments on the 12-Hour Schedule

In spite of the extra work for supervision, employee discipline and reliability usually improves after switching to the 12-hour system. A period of learning occurs by all when it is seen that discipline is absolutely necessary. Supervisors learn the weaknesses of the new system and design new ways to compensate.

In summary, employees develop a new attitude and discipline to accomplish their 12-hour workday. This occurs in a friendlier framework of more recuperation days for the internal clock to get back to normal. **And that is what shiftwork is really about ...** developing a unique, workable posture toward scheduling and the world of work. So, we have the extra-achiever feeling better with a 12-hour workday and with better stretches of work and off time for his/her extra-achievement activities!

## VERY RAPID ROTATING SCHEDULES

A solution that's becoming more and more popular in Europe is the "very rapid rotation system."

On this schedule, a maximum of two or three night shifts are worked in a row. The theory is that the internal clock doesn't have time to get out of whack. There is no long-term jet lag.

Our U.S. Air Traffic Controllers work a version of this called the "Rattler" (or the "hurry back") The Controllers' only major complaint is that they must "keep a calendar with them" to know when they must report to work.

Look at this schedule and you'll see why! Basically, it is two evenings, two days and one graveyard with two hurry-backs. This is one variation of the Controller's five 8-hour day schedule. Notice the 80-hour break that completes the weekly 168-hour cycle:

### Air Traffic Controller Schedule

**= 2 Evenings + 2 Days + 1 Graveyard (includes 2 hurry-backs)**

| DAY | WORK HRS. | OFF HOURS |
|---|---|---|
| 1 | 1600-2400 | 14 |
| 2 | 1400-2200 | 10 |
| 3 | 0800-1600 | 14 |
| 4 | 0600-1400 | 10 |
| 5 | 0000-0800 | 80 |

The main problem with this schedule is alertness on the one night shift per week. Drs. Carlton Melton and Robert Bartanowicz tell us that management promotes several things to optimize alertness during the one night:

- Management schedules a minimum of personnel and a maximum of activities on the one work night in order to keep things moving and maintain alertness during the Danger Zone.
- A variety of work is available for each Controller.
- Social interactions throughout the night are plentiful.
- Plenty of breaks are provided.[17]

The Air Controller rapid rotating schedules continue to be monitored and studied by the FAA. Insiders tell us of stressful work conditions, mandatory breaks, and poor snack or meal availability.

### MANAGING TEAM MEMBER MIX

Are you a supervisor reading this at 10 p.m.? If you are, there's a good chance that you are worried about your team that is out there tonight working without you. We trust that you have at least one "owl" in your work team.

Managing team member mix is an art form that can get lost as our plants, departments, and work groups get ever larger. We are seeing more and more scheduling specialists who do not have enough personal knowledge of the individual team members to safely schedule them! How can this be a problem?

Just imagine that you are a business passenger on a 747 night-flying from San Francisco to Tokyo. As the evening progresses, you see glimpses of each member of the cockpit crew. All of a sudden, it dawns on you ... the entire crew is composed of "larks." There's not a single night owl in the cockpit and they're looking sleepy! Feel a little threatened?

According to Richard Coleman, larks are individuals that function best by getting up early and going to bed early. Owls thrive by working late shifts.[18] The rest of us "birds" are in the middle. The lopsided scene of all owls or all larks happens all too often in our larger chemical companies, refineries, hospitals, and police patrols. Team member mix becomes lost in a sea of anonymous faces as schedulers of several hundred employees assign teams without regard to 24-hour alertness characteristics.

Managing Team
Member Mix

2 Owls
1 Lark
3 Ambi-fowl

### SUMMARIZING SHIFT SCHEDULING

Although you may never achieve the "perfect schedule," you and your shiftworkers can continuously optimize their scheduling. Whether you are running "eights," twelves," or something in the middle, the schedule and 24-hour conditions need constant attention and improvements.

We highly recommend continuous brainstorming and touch-ups to your scheduling. Since schedule formats directly impact the employees — their alertness and performance, this is not a case where outside experts and specialists know what is best for each location. Although experts are good references and advisors, we suggest that your employees continuously involve themselves in brainstorming incremental schedule improvements.

The consultant can best be used as a third-party assessor of current conditions, employee observations and safety issues, and then make recommendations. Each facility needs a champion. Each facility must also realize that a new schedule needs new training to make it work! Entering a new schedule with old habits can negate your best efforts and prove very disappointing for all.

## Chapter Four Endnotes

[1]From: *Shiftwork — the Social, Psychological, and Physical Consequences* by Paul E. Mott, Floyd C. Mann, Quin McLoughlin, and Donald P. Warwick, Ann Arbor, University of Michigan Press, 1965.

[2]From: *Wide Awake at 3:00 A.M. — by Choice or by Chance?* by Richard M. Coleman, W. H. Freeman and Company, New York, 1986.

[3]Dr. Charles Czeisler tells us that improving shiftworker's schedules can improve productivity by 25 percent and reduce fatigue-related accidents by 40 percent. From: various press releases and reports.

[4]Richey is a former shiftworker and foreman from Salt Lake Minerals and Chemical Corporation who started Shiftwork Technology Inc. of Ogden Utah. From: A New Clock for Shiftworkers — The 12-hour day and one break of seven days can cure shiftwork blues by Michael A. Verespej, *Industry Week*, April 2, 1990.

[5]Coleman gives a good account of how it actually "returned" in the form of a 42-hour per week schedule. In earlier times, it had been an 84-hour per week schedule! Now, four employees were assigned to each job rather than two. From: *Wide Awake at 3:00 A.M. — by Choice or by Chance?* by Richard M. Coleman, W. H. Freeman and Company, New York, 1986.

[6]See: A New Clock for Shiftworkers — The 12-hour day and one break of seven days can cure shiftwork blues by Michael A. Verespej, *Industry Week*, April 2, 1990.

[7]From: Report to the American Petroleum Institute by Dan J. Moller of Exxon, October, 1979.

[8]Peggy Westfall's conversations with refinery shiftworkers and their managers, 1991.

Also, Glenn McBride's first exposure to achieving a 12-hour schedule was a three-year process from the time he first proposed it as a plant technician, to the time it resurfaced, underwent close scrutiny, won approval, and was implemented in a trial run.

[9]After the first trial period, managers at many sites find themselves with essentially 100 percent of employees voting to continue the 12-hour schedule and 100 percent of the supervisors voting to discontinue it!

[10]From recent conversations with AMAX, this condition seems to have continued consistently for some time.

Employees at this award-winning site in Gillette, Wyoming have developed an excellent safety record on the 12 hour schedule. A progressive management system encourages them to take breaks when necessary and napping often occurs during their two 25-minute breaks. Also, cross-training in various jobs allows variety and breaks up monotony.

[11]See: A New Clock for Shiftworkers — The 12-hour day and one break of seven days can cure shiftwork blues by Michael A. Verespej, *Industry Week*, April 2, 1990.

[12]See: A New Clock for Shiftworkers — The 12-hour day and one break of seven days can cure shiftwork blues by Michael A. Verespej, *Industry Week*, April 2, 1990.

[13]From: A New Clock for Shiftworkers — The 12-hour day and one break of seven days can cure shiftwork blues by Michael A. Verespej, *Industry Week*, April 2, 1990.

[14]We see no problem in *requiring* employees to take vacation days during day cycles.

This is appreciated by those that relieve them! In fact, such rules often become de facto standards due to peer pressure among the employees.

[15]If you're having a different experience, we'd like to hear from you. Check for our phone number and address in the back of this manual.

[16]From: A New Clock for Shiftworkers — The 12-hour day and one break of seven days can cure shiftwork blues by Michael A. Verespej, *Industry Week*, April 2, 1990.

[17]From: *Biological Rhythms and Rotating Shiftwork — Some Considerations for Air Traffic Controllers and Managers* by Carlton E. Melton, Ph. D. and Robert S. Bartanowicz, Ed. D., Air Traffic Plans and Requirements Service, U.S. Dept. of Transportation, Office of Aviation Medicine, FAA, 1986.

[18]Richard Coleman says that about 10 percent of shiftworkers are night owls, another 10 percent are "larks." From: *Wide Awake at 3:00 A.M. — by Choice or by Chance?* by Richard M. Coleman, W. H. Freeman and Company, New York, 1986.

# CHAPTER FIVE

## F.A.S.T. Tracking Techniques

### OBJECTIVES OF CHAPTER FIVE

After completing this chapter you will be able to recommend techniques for 24-hour performers.

***Fitness***

***Alertness***

***Sleep and Sleeping Environment***

***Time off***

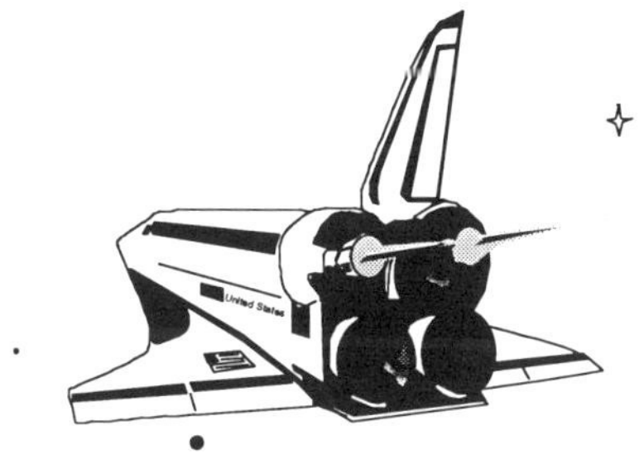

Learning these F.A.S.T. Tracking techniques will prepare you for the next chapter that will describe Performance Tracking and Improvement — where the techniques can be tracked and measured.

### U.S. SPACE PROGRAM IS THE BEGINNING

The U.S. Space Program accomplishments are a prime source for shiftwork performance improvement. The underlying F.A.S.T. Tracking techniques came from a combination of NASA procedures and policies ... and the best recommendations of safety, sleep, scheduling, health, and lifestyle experts, and our own 24-hour expert employee reports.

At the beginning of our research, we had taken the developer's standard route ... contacting sleep, safety, and scheduling specialists and looking at off-the-shelf training resources. To our dismay, they took us to the darkside. Also, many of the experts were so closely associated with sleep disorder clinics that it seemed assumed you were going to have sleep disorders working shiftwork.

Looking for the best and who was doing it well, pointed us right back to experiments and practices that had been tried or sponsored by NASA. It turned out that many of the shiftwork riddles had been solved at the Johnson Space Center in Houston, Texas.

**Origins of Shiftwork Performance Improvement**

- U.S. Space Program
- Safety Experts
- Sleep Experts
- Scheduling Experts
- Health Experts
- Lifestyle Experts
- Expert Employees

**NASA Breaks the Shackles**

As you may well agree, the corporate world tends to get bogged down with out-of-date behaviors, cultures and techniques that we can't shuck. However, a newly formed NASA, given the technology objective of the century by a young U.S. President — working under the gun of a tight schedule — and performing in front of worldwide televised publicity, broke the shackles of slow-change behavior and entered a strong period of continuous performance improvement for their NASA astronaut personnel.

Individual employee performance improvement in America has too often been focused exclusively on corporate low-scorers (who have an obvious improvement need compared to other employees). Or, personal development training was designed for management or supervisory personnel. Front-line operations, very seldomly achieved "softskill" training. Considering "alertness or performance" training as "softskill" versus safety leaves us gritting our teeth, but in most environments, it is considered so.

The NASA astronaut effort developed into what must be called true performance schedule optimization at the individual performer level. The impact of fatigue and jet lag and resulting human error in space travel drove this necessity to a top priority!

**Background of Astronauts as Shiftworkers**

The NASA shiftwork story began during the Apollo flights when it quickly became apparent that the astronauts were experiencing tremendous shiftwork-type problems. It was something other than a free ride leading up to that moment when astronaut Neil Armstrong announced to the world that the Eagle had landed on the moon.[1]

The Apollo flights leading up to the moon landings were difficult on the crews. NASA relates that after the first couple flights, the Apollo crew members, limp and weak as jello, were "peeled" out of the capsules upon re-entry. They were anything but energetic during their post-flight debriefings.

We learned one story that should be retold by every safety supervisor and scheduler in the country to their managers and employees:

**Slipping Around The Moon!**

The early Apollo flights had had several close-calls. During the early space flights, shiftwork technology (read: reducing fatigue and jet lag) had not been solved. The first Apollo craft to circle the moon almost didn't make it back because of a human error. Due to heavy fatigue and a fatigue-related slip, the astronauts accidentally erased much of the on-board computer program. It took several frantic hours of testing to assure that they would be able to return to earth, using the remaining program.

**Continuous Performance Improvement at NASA**

Through the post-flight debriefings of astronauts, much was soon learned at NASA. Scheduling, job performance, and general conditions got a little better on each space flight as jet lag, fatigue and their related shiftwork set-ups melted away through better scheduling, exercise, diet, and sleep. Performance was optimized as these improvements were implemented. NASA soon built a new brand of shiftwork that is now spinning-off to our own shiftworkers ... the shiftwork techniques learned by traveling in space for days in a small confined vehicle — with no time cues.

Another concern we heard with NASA is communications problems, jetlag with the arrival of international astronauts and their synchronization to the 12-hour shuttle shifts.

Due to the scheduling, training, and performance improvements during and after the Apollo flights, the current shuttle astronauts have overcome most of their jet lag and fatigue problems in spite of traveling 17,500 miles per hour. They no longer return home like jello and yet, they fly basically drug free! There is certainly "24-hour gold" ... for all of us with employees trying to get home in the morning in better than a jello condition.

**Incremental Improvements Pay Off!**

The NASA accomplishment is proof that even in the most extreme situations, our shiftworkers can make incremental improvements in work, sleep, and fitness behaviors that will allow them to excel beyond anything in the past. Our 24-hour employees everywhere can follow the lead NASA managers and astronauts ... and experiment, track, and hone their job and lifestyle performance.

We labeled what we have learned from NASA and other experts, the "F.A.S.T. Tracking System."[2] By getting into the spirit and procedures of such a system, your employees will be able to develop their own F.A.S.T. Track, driving improved safety and productivity for themselves as well as for your company!

## F.A.S.T. Tracking Techniques for Shiftworkers

Now we will introduce the F.A.S.T. Tracking techniques that are featured in our training workshops. The F.A.S.T. Tracking format settled on us while NASA personnel were showing us the shuttle tracking screen at Johnson Space Center Mission Control. You've seen it on television newscasts.

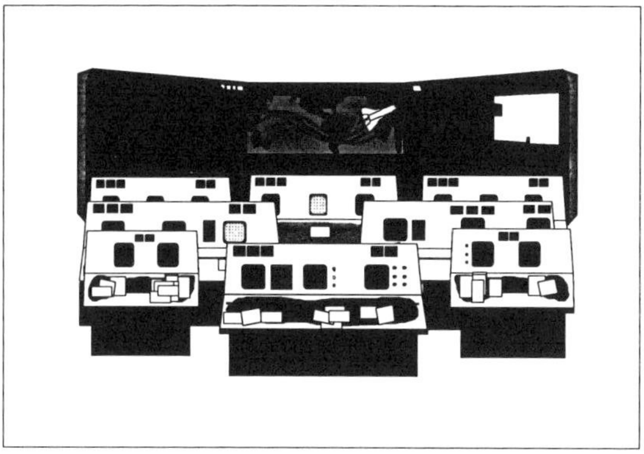

Wide and colorful well-lit screens across the back of the room and a multitude of manned computer terminals filling the center of the room. F.A.S.T. Tracking represents the fast track of the shuttle as it orbits the earth near the equator. It also represents the employee's tracking (charting) their efforts in the four performance areas ... Fitness, Alertness, Sleep, and Time off.

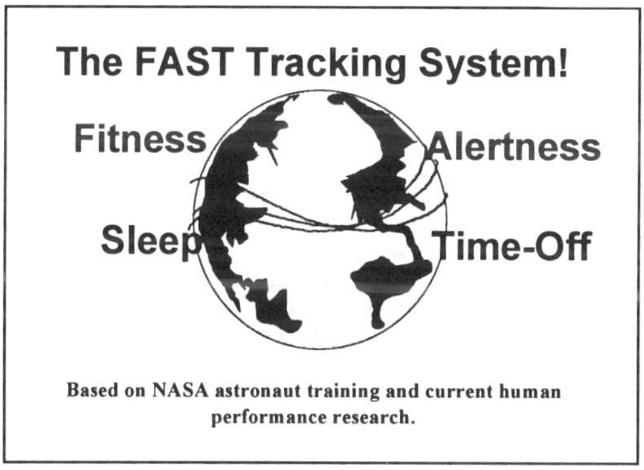


One secret of F.A.S.T. Tracking is that it is a very interrelated system. If the employee improves efforts in just one F.A.S.T. Tracking quadrant, he/she will soon notice a definite overall performance improvement. Family members will often notice it, too. Change is difficult for some and that is why we always recommend training to support such a motivational tool such as F.A.S.T. Tracking.

Let's say employees make an incremental improvement in their sleep schedule by eliminating alcohol near bedtime.[3] They will soon find that they have broken the pattern of waking up after two hours into their long sleep. They may express amazement that the other three areas have been affected as well — alertness on-shift has improved, willingness to exercise has improved, and so has the quality of the individual's time off. He/she may be less moody around the house and more inclined to participate in family activities.

## Fitness Techniques

In F.A.S.T. Tracking, we explain fitness as depending mainly upon diet and exercise. Being physically fit does not mean having a poster body or aerobic instructor. Being fit means having a strong cardiovascular system and good flexibility. An important lesson: It is better to be actively staying fit than just having a great-looking body and being inactive. The process is nearly more important than the outcome. More on this when we get to Performance Tracking and efforts.

### *Diets at NASA*

We learned some things by studying the astronaut diet. The Space Shuttle orbiter carries seventy-five different kinds of food and twenty different drinks during each mission. The food is packed in moderate quantities ... about 3000 calories per day for each astronaut. Experts say that we earth-bound types consume an additional 1000 calories than we admit. Accordingly, the 3000 astronaut calories seem to be in-line with many diets of active healthy people.

Meals on the shuttle are planned carefully and on the lighter side. Lighter in quantity, fat and sugar-content, healthy meals promote top performance and alertness during the shift. Heavy, fatty meals can be dangerous in this critical job.

What we can practically learn from NASA is that eating lighter meals on a fairly consistent schedule improves alertness and performance on the job. Due to the impact of food on alertness during the "Danger Zone," lighter meals can be more than just healthy — lighter meals can mean improved response time, or more alertness on-shift or while driving home. This can mean less occurrences of human error, our goal.

### *The F.A.S.T. Tracking Diet*

When Peggy talks about the F.A.S.T. Tracking diet in her shiftwork training workshops, she hits home 100% of the time when she pulls out her favorite "shiftworker" lunch kit — a 24-inch heavy duty tool box! She takes each item out of the box and explains to her audience:

"This is a typical shiftwork lunch. Remember, I was married to one of you for almost 15 years! In fact, my job, as the spouse of a refinery shiftworker, was to ensure that the proper ... no, make that ... the essential items ... were included. Okay, mind you, my ex-husband was a big guy about 6 foot two. On his way out the door from our home, he would pick the lunch kit up and weigh it in his hand. If it felt full, he would proceed out the door. If not, he would take the time to check it out, making sure the necessary items were included. He ALWAYS caught the missing items with the raise of an eyebrow. I would feel like I let him down.

"First, a light snack of a powdered donuts to have with coffee when he gets to work. Oh sure, he'll share them with his buddies, there was no way he could eat 24 donuts by himself. Right? Two hours later, time for the main meal. How about a 12" meatball sub! Of course, no meal is complete without a liter of soda — no diet-stuff here, and a 12 oz. bag of nacho chips and a can of bean dip. And, then there's 2 or 3 slices of leftover pecan pie for dessert. Well, by this time, I bet he's about ready to go outside and take readings ... better take a snack. (She pulls out a jumbo Butterfinger bar.) Before the shift was over, there was still time for a small snack to keep him going ... a honey bun warmed in the microwave, yummm!

"Now, this meal was designed for an 8-hour shift on days. Along with his main cuisine, there would be two packs of real cigarettes — no light versions, a can of Skoal, and a roll of antacids. Might also add a few cookies in case he got hungry later. On graveyards, he might still take the meatball sub, but he'd also include 4 eggs, 1/2 a pound of bacon, some sausage if we had any around, and a bag of tater tots.

"How does this lunch look to you? Familiar?"

Nods and affirmations follow. She usually fields requests for the contents before the end of the session! Knowing what she does now, of course, the meal would be packed differently. Efforts to eat to maintain alertness would be geared towards performance foods.

Lighter meals are necessary! Some shiftworkers feel that since they are so "busy," they need heavier meals like the one above. How many times have you heard the office paper shuffler talk of his busy day as he heads to the chicken-fried steak line in the cafeteria. Just because we are busy, doesn't mean we should eat heavy food. First, determine what your "busy" is. Are you answering the phones? Are you mentally-stressed? Or, are you tired from loading boxes to a flatbed? There are different types of fatigue, different types of "busy."

Along with this thinking is carbo-loading. Well, unless your shiftworkers are in physically-demanding jobs for their full shifts, loading up on carbohydrates may give them the opposite result. If duties are limited and they are bored, they may become overwhelmed by feelings of fullness, calm and fatigue.

Many shiftwork jobs are combination jobs: monitoring equipment, making adjustments, judging flows/speed/levels, and occasionally bursts of quick energy to take readings. Some jobs such as law enforcement and emergency medical services require workers to go from one to the other. So, just how does one formulate lighter *Performance Improvement* meals?

Try beginning the day with low fat and high proteins meal. Maybe a "breakfast" meal would include the low-fat egg product scrambled with lean turkey with a bagel and piece of fruit on the side. Use a non-stick spray on your pan to prevent sticking. Other foods great for performance include fish, lean beef, chicken (without skin) and all sorts of vegetables.

Foods that would make for good snacking on-shift include popcorn — cakes, air-popped or other forms. Watch the packaged microwave popcorn because many are loaded with fat. Make sure your vending machines are carrying the lighter versions of popcorn and microwave foods. Some shiftworkers are also avid hunters on their time off. You'll find deer jerkie as a popular snack food in south Texas chemical plants.

We recommend boiling or broiling your foods rather than frying. The diet should gradually progress to carbohydrates by the end of the shift or before bedtime. After midnight, cut back on eating. Stomach distress is the usual midnight complaint. Lighter, healthy meals improve the day person's fitness in the long run ... lighter meals affect the shiftworker's performance on that very work day!

Your company's dietitian or hygienist will probably be glad to help you and your shiftworkers with lighter, low fat performance-based recipes and menus. One shiftwork-related reference you should review is *Managing Your Mind and Mood Through Food* by Dr. Judith J. Wurtman. The original 275 page book was published in 1986 by Rawson Associates.[4] She has written other more recent books on diet and the impact on mood, brain chemical levels and behavior. You'll find her work extremely interesting and relevant to performance in 24-hour operations.

Notice that we aren't promoting eating a diet solely of rabbit food! Many shiftworkers will tell you that they can't make it through the night on leaves, seeds and bark. Many workers will be sensitive to protein diets or carbohydrate diets. The recommendation we are making is to bring diet and the impact on human performance to their attention. Promote the concept that "what you eat" today determines more than their physical condition 20 years from now ... it can make the difference in their ability to drive, stay awake and make critical decisions today!

Share the current research and provide knowledge to your 24-hour employees.

### *Exercise at NASA*

Now for exercise. Do you know how much routine exercise is required of the shuttle crew each day while in flight? Many of our trainees are surprised that astronauts even exercise at all on space missions.

Many missions include probes to monitor heartbeat, sleep and exercise patterns. You've probably seen the astronauts in orbit exercise on television, strapping themselves to a bungee cord treadmill.

Well, our answer to the leading question is: at least an hour per workday of this indoor exercise helps maintain their fitness, bone and muscle mass and heart strength while in space. This exercise also raises their heart levels to help set their circadian cycles to the appropriate work shift. In a weightless environment, some real fitness and health problems have appeared. For instance, we know that an astronaut's heart will noticeably shrink without the daily exercise. With longer missions and Space Station plans, the physical losses become more apparent. Be sure to follow future flight itineraries (NASA Internet pages) as the circadian specialists test and apply their knowledge to improve these conditions.

There is plenty of backup for NASA's stance on crew exercise and we haven't found a lot of space potatoes in their program. Fitness experts say that even moderate exercise done in short bursts several times each day will make all of us healthier.

One such announcement was at the 1992 International Conference on Physical Activity, Fitness and Health in Toronto. Dr. Claude Bouchard, professor of exercise physiology at Laval University in Quebec City, organizer of the conference, says that this view has not been commonly accepted. Even a few spurts of mild exercise each day speeds up metabolism and improves the body's health. These "spurts" can be performed on a control center treadmill, stairs to the second floor, or a brisk inspection walk around the plant, or patrol walk around a park. Exercise increases the flow of energy in the body and that impacts directly on the fat component, according to Bouchard.[5]

There are many plants that have put treadmills and weight machines in their control rooms. Sometimes, the equipment is used as a coat rack, but sometimes personnel actually use it! Promoting your wellness efforts as performance can have a dramatic effect on your "fitness" program. Friendly competition ... control center vs. control center can also motivate operators from their comfortable, dark, hypnotic control room chairs to the treadmill for a quick sprint to increase their heart rates!

The price of not exercising has become a major health issue these days. Analysts state that today's average costs of company health insurance is a whopping 24 percent of after-tax profits.[6] When we work with clients who are making shift schedule changes, we also recommend that they implement their F.A.S.T. Tracking program at the same time. It is important to teach employees how to work the new shift schedule, what types of meals to pack, and how fitness affects their alertness. Companies that have implemented a F.A.S.T. Tracking program at the same time have found a pink-cloud effect spilling over ... right into their health and fitness program. Here is a valuable piece of motivation that you can build on if you give it the right spin.

### *F.A.S.T. Tracking Exercise*

For 24-hour people, the time of exercise makes a great difference. Experts tell us that the best time to exercise is within an hour of awakening. Exercise can help employees get moving, raise body temperature, and set their "internal clock." The internal clock controls when we want to sleep and our times of maximum alertness. As we know, exercising increases employee energy, self-perception and motivation, making them more likely to speak-up and intervene.

A common report is, "there's no time to exercise on 12s." Anyone who works away from home and has kids knows this is true! BUT, if safety and alertness are to be given a priority, then alternative methods of exercising, including on-shift time blocks, should be considered. The control room exercise center is a great idea. Trained by sports trainers and nutritionists, promoted as performance vs. wellness, employees can incorporate exercise with their other duties.

***Stimulants and Depressants***

Okay, what about stimulants and depressants for 24-hour people? Do you visualize a lot of drugs being used by astronauts? We did. However, we found just the opposite, except for an emergency kit. The astronauts work, go to sleep, and wake up basically drug-free, while zipping around the globe in an adventurous environment more hostile than even found in Robert L. Stevenson's Treasure Island!

**Yo-Ho-Ho ...**
**on a Pre-Space Journey**

"Fifteen men on the Dead Man's Chest -
Yo-ho-ho and a bottle of rum!
Drink and the devil had done for the rest -
Yo-ho-ho, and a bottle of rum!"

They certainly have eliminated the old rum problem.[7]

Back on earth, we haven't found a single expert that recommends drugs as shiftwork performance aids! Sleeping pills, alcohol, and tobacco should be absolutely minimized in our critical 24-hour jobs. Coffee may be used but shiftworkers should watch for sleep problems and interruptions from the caffeine. Most sleep disorder specialists advise to stop drinking coffee at least four hours before bedtime. Alcoholic beverages should not be enjoyed right before sleep either as they may interrupt the sleep cycle.

As of this writing, over the counter sleep aids with the hormone melatonin should be avoided until clear data supports its safe short and long-term use. Shiftworkers should consider their mental and physical condition should there be a call-out or family emergency midway through their sleep cycle. Being under the influence of sleep-inducing medications may have "hangover effects" impairing response time and alertness the next day. Individuals who are experiencing sleep-related problems should contact their physician for assistance. If medication is prescribed, it should be determined if the individual will experience any effect of the drug the next day.

***Summary — F.A.S.T. Tracking Fitness***

Okay, let's summarize for you how a shiftworker can F.A.S.T. Track fitness and health ... remember, these are the critical items that must be closely controlled by the employee.

**CRITICAL Fitness and Health Efforts**

- •Light meals on work days
- •Consistent eating schedule
- •Exercise one hour per day (min.)
- •Minimize drugs, coffee, tobacco, alcohol

Shiftworkers can improve their fitness and synchronize their internal clocks at the same time.

Remember that NASA astronauts fly drug free. NASA believes that a highly trained, well-rested human can out-perform a human under the influence of drugs. The 24-hour person must minimize any drugs, tobacco, or alcohol in order to F.A.S.T. Track.

## Alertness Techniques

Remember that the Danger Zone statistically occurs in the late afternoon and very early morning hours. The individual goes into a lower alertness and slower response state for two or three hours. If things get really tough, he may experience a Micro-sleep Slip, that nodding off for a few seconds at a time, while driving, monitoring, talking, making critical judgments.

The early morning Danger Zone is the time that an extraordinary percentage of all types of slips occur ... by truck drivers, surgeons, pharmacists, telephone operators, and postal workers! Many of these slips break into the news as safety incidents or fatalities. Many of these slips never get reported and repeat themselves until an event happens.

### *Low Alertness and Early Morning Incidents*

The greatest rate of shiftwork slips occurs during the Danger Zone after midnight. Especially, from 3 a.m. to 5 a.m. According to sleep researchers, after midnight, an employee's reaction time can nearly double.[8] If you add lack of sleep, no exercise, poor diet or recent heavy meal, this reaction time slows even more. He may fail to spot a problem (Attention Slip) or react slowly to an alarm.

When a person mis-operates, mis-judges, reacts too slowly or falls into micro-sleep for a few seconds, an explosion, a plane crash or a highway accident can be the immediate fatal result. Chernobyl, Three-Mile Island, Bhopal, the Exxon Valdez ... all teach us the lesson of reduced alertness and the late night Danger Zone.

### *Working During the Danger Zone*

Working during the statistical Danger Zone is necessary for many of us. As we have seen, fitness will contribute to alertness and response-readiness. Ideally, alertness can be physiologically improved with better shift scheduling and sleep/diet/exercise habits. Countermeasures are needed if micro-sleep occurs. Even in the best of environments, we as humans tend to get occasionally sleepy while working or driving. Increasing your employees knowledge of these countermeasures can improve their handling of the next sleepy event.

The Danger Zone is where the effects of being actively physically fit will show directly. If you feel sleepy, attempt to run up and down some steps, or run in place to get your heart rate going again. If you're not used to exercising, you will be less likely to make this effort.

It seems that motivating oneself to run up and down those steps while sleepy is the biggest problem. Heck, if you felt like running up and down steps in the first place, you wouldn't be tired, would you? Well, being aware of the onset of sleepliness is the first step to taking the countermeasures. Be sure to have your safety team discuss fatigue and the signals. Teach them the countermeasures and suggest team intervention!

When drifting into micro-sleep, immediately change pace. Often, when micro-sleep occurs, the individual is not able to make the judgment to take a break or step outside. More often than not, the individual will attempt to continue to drive home, nodding off several more times until the vehicle runs off the road, or the control room door slams, or a horn blares. In many of Peggy's workshops, individuals will frequently comment that driving faster helps keep them awake. We DO NOT recommend this! It's scary to think of these fatigued individuals racing to get home sooner at the same time our children are traveling to school. We do recommend that sleepy drivers stop their vehicles in a safe location, and get out of the car. Walking around may help, along with some conversation.

Interestingly, driver loneliness can bring on sleepiness! How many times have you been the designated alert driver with a car full of sleeping people and found yourself dozing off! Always try to keep one person awake with you. If this isn't possible, listen to an interesting talk show while driving, or get your C.B. radio out of the basement and hook it up. Long distance drivers are also known to be tape-listeners, brushing up on condensed book tapes, language tapes or educational classes.

Remember, the employee needs to change pace and take breaks often. Giving Danger Zone employees an extra break on-shift to regain alertness could save lives and many dollars.

Newer multi-media computer-based training can prove to be useful to promote alertness on-shift. Many developers now incorporate the use of sound and music to keep learners interested. Williams Learning Network in Houston, Texas, is moving into a server mode of delivering their whole technical training library! This is a great opportunity to do 24-hour a day training on-demand AND at the same time, keep employee's minds moving and thinking, optimizing off-peak shift hours.

Almost all employees already snack to maintain alertness. Ed, a respected refinery Shift Foreman mentions that his crew utilizes low-fat popcorn as a performance food during the wee hours. Just the smell of popcorn cooking wakes them up! Be sure to review the fitness portion of F.A.S.T. Tracking for more ideas.

Napping is proving effective in industries where it is allowed. (See the cockpit napping by the FAA and also the experiences of AMAX Coal Company. James C. Miller has recently presented a paper to SAE suggesting naps for truck drivers operating at night.) When mentioned during workshops, attendees say, "It'll never happen at XXX Co.!"

### *Compensating During the Danger Zone*

When working in the Danger Zone, there are critical procedures the employee should perform to compensate for the reduced alertness. Stopping and thinking twice before making important decisions is necessary to prevent Attention Slips. The employee should move carefully and deliberately due to the higher possibility of making a Fumbling Slip. A good example is a critical procedure being performed at 3:30 a.m. The employee stops and thinks the process through before acting on it, even discussing it with a team member. Using checklists and procedures should be performed routinely. If checklists and procedures are used in the employee's normal job, they should become the "bible" during Danger Zone work. Motivation to use checklists and to "make rounds" is often the first thing to go![9] Unfortunately, omitting checklists leads to Attention Slips — forgetting important steps. Fatigue is a major factor in skipping steps to get a task finished sooner.

As alertness levels decrease, so may the individuals ability and desire to complete all of the steps of dangerous procedures. Thus, you'll be more likely to see the best trained employees skipping steps on graveyards while fatigued.

Another serious problem is that older employees who are training younger employees may train them according to these shortcuts on graveyards. The night shift may do things differently than the day shift. The new employee may speak-up at first at this difference in procedures, but could eventually fall into the group-think due to team pressure or fatigue.

### *Job Activity Extremes Among Continuous Process Operators*

We see two job activity extremes in oil and chemical plant operators. In the 3 a.m. control room, monotony becomes the rule in many well-run plants. Staying alert when there are no surprises or emergencies can challenge the best Board Operator.[10]

On the other hand, the Outside Operators often face heavy, even aerobic activities during the wee hours. If you've ever performed a training needs analysis for an Outside Operator, you know what we mean! They are the ones running up and down flights of stairs on shift taking readings, making adjustments, checking flows. Try following the Outside Operator around for a few nights in a row and you'll see how exhausting their jobs can be. The Board Operator's job is quite different from the Outside Operator's. Experienced employees recommend the following on-the-job actions.

**Working in the Danger Zone**

- Get fit
- Take breaks often
- Change pace
- Eat light snacks
- Stop and think twice
- Use checklists

### *Using Teamwork to Maintain Alertness*

NASA and the FAA are experimenting with ways to monitor crew member alertness using cockpit instruments. Most of our plants haven't gotten into that advanced mode yet, however there are devices on the market that can be used to measure the employee's speed of response and compare it to the employees "norm." This device can be used to police employees for fatigue, alcohol or drug use. Since it is not physically invasive, it will not show which effect is being tested. However, once a benchmark has been established and the employee tests outside of that benchmark, a flag will be raised. The optimum use of these devices would be as a motivational and tracking tool used by the employees to test their attempts at improving alertness. There are other methods that teams can employ to monitor and improve alertness.

### *The A.Q. Test*

The A.Q. (Alertness Quotient) Test is a simple device that teams can develop. This is simply monitoring each other and testing with the question, "How alert do you appear to me?"

### *The Alertness Pact*

How do you get there from here? Discussion with or training team members is the ONLY way! Present the facts on fatigue and circadian rhythms and discuss proactive solutions. One of which is the Alertness Pact. Undoubtedly, they *will* make fun of the A.Q. during your session and while practicing it on each other. But, if things go well, a piece of the training will stick. If you hang around the control room at 4:30 a.m., you might hear one operator suggest to another, "Hey, you look horrible! We've got another hour to go, why don't you go outside and wake-up!" or "Hey, Stella, didn't you get any sleep yesterday? Why don't you let me watch the board while you fix us a pot of coffee!" Alertness Pacts are made between team members to monitor each other and "A.Q. Test" each other often during the Danger Zone. With a little boost in the right direction, good-hearted team members will have no trouble making suggestions and keeping each other wide awake!

Such a simple thing to do, but this simple team monitoring is often omitted by shiftworkers, especially new teams! Quick intervention by a team member, whether in a control room, in a vehicle, in a cockpit, can prevent a serious accident from happening. Team pressure can prevent someone from speaking up. Discussing these methods as part of your safety program may do wonders for your team coordination.

Some teams take advantage of the inside/outside job activity extremes. Every time an Outside Operator returns to the control room, it is his/her task to *"hold a reveille"* or do whatever is needed to rev up things. The Control Operator and any Outside Operators inside at the time will be forced to come to life.[11] We know of a few Operators who enjoy coming into the control room SLAMMING the door to wake up the crew!

### *Commuting Home*

Although not all shiftworkers are aware of the Danger Zone, many have expressed great concern over their drive home at midnight, 6 a.m., 7 a.m. or 8 a.m.[12]

Such a concern is well-founded. Statistically, shiftworkers experience twice as many motor vehicle accidents as those who work the day shift.[13]

### *Commuting Home at NASA*

The Shuttle astronauts experience the most unique trip home of any shiftworkers we've found. They re-enter the atmosphere upside down and backwards! They then swap ends of this space age glider and fly to a perfect touch-down on the desert floor! However, it may be easier for astronauts to land the space shuttle than for your shiftworker to return home in his/her land shuttle. This is because the Shuttle crew performs re-entry and touchdown only when the entire landing team has had plenty of sleep. If the Shuttle team has stayed awake too long, they maintain orbit until they have slept! How does that compare to your current standards?

The trip home after working nights can be a very dangerous trip and a real set-up. This is a drive where your shiftworkers need to take extra precautions. Some driving tips shiftworkers have given us are the following:

**Commuting Home Tips**

- Take something to drink, cola and ice, decaff-coffee.
- Exercise briefly before you get in your car.
- Stay in touch with fellow shiftworkers by CB or cellular phone.
- Drive very defensively.
- Watch speed limits, don't try to hurry home because you're sleepy.
- Stop and take a nap or pull into a quick-stop if it's safe.
- Start a van pool or car pool with someone you don't like!

### *Summary — F.A.S.T. Tracking Alertness*

Let's summarize the alertness chapter:

- On the job ... change pace and take breaks often. Eat lighter and eat light snacks as refreshments during the shift. Step outside occasionally, if possible.
- During the Danger Zone or if experiencing jet lag ... stop and think twice before making any important decision. Also, really use checklists and procedures during this time.

- Have team members monitor each other with the A.Q. Pact.
- Finally, on the drive home, stay alert and stop when necessary!

## Sleep Techniques

### *A Heavy Price for Not Sleeping!*

Remember the $70 billion paid annually in the U.S. for lost productivity, industrial accidents, and higher medical bills due to reduced alertness?

The Better Sleep Council attributes much of this alertness reduction to disrupted sleep and sleep disorders.[14] Dr. William C. Dement states that millions of Americans generally do not get enough sleep ... due to socializing, television, and the general fast pace of the country. In fact, they build up a chronic sleep debt.[15]

Dr. Ann Romaker says that there is a growing interest in sleep deprivation as a medical problem. Looking at it this way, 10 to 15% of Americans develop a sleep disorder according to Dr. Romaker. She also states that 35% of the population experiences continuing difficulty falling asleep or staying asleep. This is in addition to the thousands that tolerate constant fatigue.[16] The National Center for Health Services ranks proper length of sleep as one of the six most critical factors affecting "mortality and morbidity" in our population.[17] Dr. Timothy Monk advises supervisors of shiftworkers that sleep is as important as oxygen.[18]

Let's discuss sleep and the sleeping environment. Remember that sleep affects the other F.A.S.T. Track areas. The following information is what convinced us that NASA has moved way ahead of the rest of us in shiftwork technology.

### *Controlling Sleep at NASA*

During Space Shuttle flight planning, the Flight Activities Officer schedules everything around one astronaut activity. When we ask our shiftwork classes what they think that activity is, they usually answer that it's the work schedule or the cargo bay projects schedule.

The real answer surprised us too ... and it will surprise many of our readers. On this particular visit with the NASA F.A.O., we sat at his T-desk, an arrangement of office furniture quite popular along these hallways. Questioning him on how the 12-hour shift schedules are configured, he startled us with this revelation. We learned that all flight activities are planned around the crew sleep schedule! This is perhaps the most important thing we've learned from NASA ... a shiftworker's sleep schedule is more crucial than the work schedule! Eight-hour blocks of sleep time is the first element plotted on the schedule, along with some personal pre- and post-sleep time. After all the sleep schedules have been blocked, THEN, the work needing to be done is planned and assigned.

**The Most Important Item ...**

Work Schedule?
or
Sleep Schedule?

***Enforcing the Sleep Schedule***

The NASA astronauts must maintain high alertness and near perfect performance throughout their shifts. During flights, the Flight Activities Officer works at a console at JSC Mission Control to maintain and manage any changes to the scheduling. The daily sleep schedule of at least eight hours is highly enforced.

***It Isn't the Work Schedule!***

So, keeping this in mind, imagine the responsibility of determining how to arrange the job tasks around sleep: lift-off, shuttle maneuvers, payload operations, and re-entry. Again, it isn't the work schedule that determines the order of activities on the shuttle ... it's the sleep schedule![19]

Managers take note! This is 180 degrees off from the thinking of most shiftworkers, their supervisors, and their schedule coordinators. How are personnel scheduled at your facility? Usually, it's the job load. We are seeing improvements to scheduling occurring in many industries though, standards are being developed on safe hours to work and hours to rest. Only now, as of this writing, are human factors even being considered in the oil and chemical industry, underground mining and rail.

### *Sleep Shifting Before Launch*

Internal clocks of the astronauts are adjusted for optimum performance before the count down for lift-off at Cape Canaveral.

What we call "sleep shifting" (moving sleep — the internal clock — to a new schedule) begins up to two weeks before the launch.[20] Think about the feeling of "zero hour" for a first time astronaut. He or she may be on the "blue team" that is scheduled to go to sleep as soon as three or four hours after the blast-off.

### *Shifting Sleep Earlier*

Here's more surprising information: The maximum that the start of an astronaut sleep period may be adjusted earlier than the previous day is 1-1/2 hours.

NASA has found that it's about all the internal clock can shift in one day without developing jet lag and therefore reducing alertness and performance. Jet lag is simply getting your sleep schedule out of synch with everyone else. Almost like being in the Twilight Zone, many shiftworkers feel as if they're walking around in a daze, not able to eat or process their food well, not able to communicate, falling asleep while everyone else is energetic — or feeling jet lagged.

### *Shifting Sleep Later*

The maximum that the start of a sleep period may be shifted later than the previous day is 2 hours.

NASA prefers shifting to later sleep hours rather than earlier because it's easier on the circadian cycle. This is because humans tend to run on a 25-hour clock and therefore sleep a little later every day unless something forces them to wake up at the same time each day. Sleep schedule movement is critical! If the employee must move the sleep schedule due to a rotating work schedule, it should be shifted forward as slowly as possible over several days. NASA has found that this is necessary for top crew performance.

### *Returning From Days Off*

Employees who have been on a long break and now must return to work should manage their sleep schedule prudently. If the return is to a night shift, the employee should begin adjusting his schedule at least a day in advance of the oncoming shift. Most shiftworkers state that they take an extended nap on the day they are to come in for a night shift after being on a break. This is good, but they could go one better. If on a set of days off and the next set is a night shift, try to stay up later and later each of the days off until the night shift. This could minimize the shock of the first night's fatigue. Discussion of sleep and sleep management is important to making the F.A.S.T. Tracking program work.

Like NASA, your employees need to set up a definite sleep schedule ... setting a goal to minimize sleep shifting to three hours tops per 24 hours, two would be even better. That's not quite as strict as NASA's scheduling constraints but it should have a quite noticeable effect.

### *Maximum Crew Awake Time*

Another NASA crew scheduling constraint is the maximum 16-hour "awake time" allowed on launch day before lift off. Special "hold time" on the pad may not extend this time to more than 18 hours. This rule is highly enforced.

You can see that NASA has honed the astronaut sleep and work schedules to a precision that we can find nowhere else. The greatest travelers in our known galaxy are continuously working at preventing jet lag during travel and work! Companies and their employees need to learn from the techniques that have been pioneered by NASA.

***Controlling the Sleeping Environment***

The other sleep item that employees must control extremely well is their own sleeping environment. Have you seen the sleeping environment aboard the Space Shuttle?

The astronauts sleep strapped in a cage, often upside down! Thank goodness the head can be strapped in so that it won't bob around due to zero gravity. With so much excitement and so many activities going on, sleeping crew members must be supplied with eye shades and ear plugs. Your terrestrial employees can probably get better control over their sleeping environment than the Space Shuttle astronauts ... or can they? Think of the obstacles to "day sleeping" in a typical 21st century home. Some of the choice items that our shiftworkers mention are:

**Obstacles to Daytime Sleeping**

- Neighborhood dogs-barking-in-tandem
- Telephone ringing ... often "friendly" calls
- Children: laughing, crying, talking, playing
- Sunlight
- Lawn mowers at all hours
- Ambulance, fire, police, community sirens

Noise, light, and interruptions are the greatest problems reported by 24-hour people. Sleeping during daytime hours can be difficult. Sleep experts tell us to sleep in a dark, quiet room that mimics night time. Remember from the NASA light experiments that light can make the "internal clock" feel that it's time to wake up and go to work! Some advice:

- Use dark shades, blinds or hang heavy, lined drapes to cut light.
- Apply aluminum foil to the windows
- Use an answering machine to catch calls

We would go further than this if necessary:

- Forward calls to a relative, neighbor or spouse's pager or portable phone
- Turn off the door bell ringer during sleep hours
- Hire an answering service to screen unimportant calls

- Install a second line that's left on for emergencies

Most of our shiftworker attendees report using earplugs like the shuttle crew ... or would you believe extremely similar to the ones they use at work! Some have even sound-insulated their walls and flooring to cut down on the daytime noise. We hear of shiftworkers who sleep, while working a night shift, in the couple's travel trailer or in an above-the-garage apartment.

Temperature control of the sleeping environment is also important, especially in warmer climates. Keep the room at around 68 degrees.[21] Keeping the whole house at 68 degrees may get a bit expensive for some climates. Many have found success with the installation of a small air conditioning unit in the bedroom. Doubles as a cooling mechanism and a white noise maker, which follows.

**The Voices of Experience**

CUT LIGHT
- Dark shades or heavy curtains
- Aluminum foil on the windows

CONTROL TEMPERATURE
- Keep sleeping room around degrees

REDUCE NOISE
- Use answering machine
- Turn off the phone
- Use call forwarding
- Have spouse/friend catch
- Turn off door bell ringer.
- Hire an answering service
- Use earplugs
- Use white noise

***Using Familiar White Noise***

Prior to a launch, the astronauts listen to audio tapes of "normal" spacecraft noises. This prepares them to sleep during their flights. Shiftworkers can add white noise to their own sleep environments to cover "spouse noise," children playing, household noise, neighbors.

Fans and radios make good continuous noise devices. The National Safety Council suggests that we not use televisions as white noise, but setting the timer to help get to sleep may help. Many 24-hour people run their air conditioning units whether or not they really need them for cooling.

Peggy heard of one shiftworker with six children in a three bedroom house who ran a vacuum cleaner under his bed on graveyards! We don't recommend anything quite that aggressive! Nevertheless, moderate white noise will make other people in the home more comfortable and get them off the "eggshells" they walk on to keep from waking you. There's one minor problem with white noise. If it stops, the person sleeping will immediately awaken! One shiftworker we know of said that his wife was his white noise. If she left the house for any reason, he'd wake up in a sleepy stupor and wander around the house aimlessly looking for her!

### *Using Familiar Patterns*

Timothy Monk and other sleep experts recommend that shiftworkers practice pre-sleep "rituals" anytime they want to sleep during the day. These are the same rituals that we would use prior to going to sleep at night. Showering and brushing teeth right before going to sleep is a pattern for many of us.

### *Sleep Disorder Centers*

In spite of applying the best efforts, the entire population cannot attain a perfect sleep record. Since the book does not address sleep disorders or clinical approaches, we want to suggest some agencies that can provide this type of information. More than 170 sleep disorders centers now exist in the United States and Canada.[22]

With sophisticated diagnostic equipment and staff members trained in sleep medicine, these clinics help an estimated 85 to 90 percent of their clients sleep better.[23] Sleep disorder clinics are now listed in the telephone yellow pages. Also, most larger communities contain at least one practicing sleep specialist.. You can also try

American Sleep Disorders Association
604 Second Street S.W.
Rochester, MN 55902

National Sleep Foundation
122 South Robertson Blvd.
Suite 201
Los Angeles, CA 90048

American Narcolepsy Association
P.O. Box 1187
San Carlos, CA 95070

National Institute of Mental Health Inquiries Section
5600 Fishers Lane
Room 15C05
Rockville, MD 20857

Canadian Sleep Society
c/o Dr. Harvey Moldofsky
Western Division of the Toronto Hospital
399 Bathurst Street
Toronto, Ontario, M5T 2S8

## TIME-OFF TECHNIQUES

### *Balancing the Act!*

Michael Fortino of San Francisco is a contemporary life style — experimenting consultant who carries lap top computer, portable phone, fax, printer, copier, and modem everywhere. He tells clients "the modern working individual is like me — a juggler trying to keep both rubber and crystal balls in the air. If one of those balls that is rubber drops, it bounces," he said. He continues, "If the crystal balls of health or family drop, they permanently shatter and you can't bring them back."[24] This rings true and very loudly for the shiftworker. The shiftworker must put extra effort into balancing the juggling act. Time off is where the shiftworker's act is balanced!

### *Time off: A Picture of America*

**Dual-Earner Statistics!**

**Dual Income American Families ...**

- 25% of spouses work different shifts. (Higher percentage in younger families!)
- 50% of young, dual income families contain one or two shiftworkers ... percentage is rising!

In America, spouses work different shifts in one-quarter of all dual-earner couples![25] In the younger families, it's even more common. Half of all our young, dual income families contain one or two shiftworkers ... and this percentage is increasing!

Unfortunately, we learn from a major study of shiftwork and marriage by Lynn White and Bruce Keith that shiftwork greatly increases the probability of divorce.[26] During Shiftwork Assessments on-site, we typically find a divorce epidemic, poor morale, poor health, poor scheduling linked. Not saying one necessitates the other, but chances are, finding multiple situations where one has lead to the other.

### *Working 13 Months a Year!*

Parents have special scheduling needs. Women, in particular, in dual-earner couples, can experience an uphill battle. The data show that over the course of one year, employed females (in general) work an extra month of 24-hour days if paid jobs, housework, and child care are added into the equation.[27] Men, as heads of a single-parent households, can experience a similar overload. What a set-up for fatigue-related slips on the job![28]

***Setting That Clock!***

Any employee with a new baby will often state that he is working two shifts — one at work, one at home. On top of that, the second shift is erratic and full of wake-up calls! Richard Coleman says that infant biological clocks are poorly developed until the age of three or four months, causing a random eating and sleeping schedule.[29] Mutual role-sharing between husband and wife can help this situation. In many cases, unless help arrives when the newborn arrives, a sleep deficit for both the husband and wife will be inevitable.

***Dual Careers or Dueling Careers?***

Peggy's work with shiftworkers and their families has yielded an interesting point:

When one family member works a shiftwork job and the other a day schedule, impending conflicts and crises must be monitored and often rescheduled to a better time. Peg says the couple must decide ... will it be dual careers or dueling careers? Both spouses must stay tuned to the other's schedule (and personal Danger Zones) and figure out how each can synchronize with the other. Is this a good time to eat, play, work, or discuss?

***The Dueling Zones***

Watch for the following two types of Dueling Zones that can have one of you practically begging for a pair of pistols ... ready to turn and fire at the least upset!

- **THE DZ YOU!** You are experiencing a Danger Zone type dip and are more sensitive than usual and the spouse is in great shape. You're tired, just got off work and pretty stressed. To top it off, your spouse is frisky!

- **THE DZ SPOUSE!** The spouse is experiencing a dip and extra vulnerability and you are in great shape, ready to go here, do this, discuss that.

With two unsynchronized internal clocks in the same room, the person in best shape may jump right into a touchy subject, not realizing that the other is experiencing a DZ. They get a shock as their mate quickly "loads their psychological weapon, turns and fires" with little or no warning!

Day workers often refer to their own special Dueling Zones as "the Friday Night Fights." Both are home together, DZ'd: fatigued and grumpy. However, luckily for them, they're also synchronized on a Friday night! Since both are synchronized to the same vulnerable dip, they often realize that they are too fatigued (I'm tired ... so am I) to discuss the matter rationally and laugh it off or delay the debate to a better time.

### *A Little Dueling Now and Then!*

A small blow-up now and then between married couples may be good for the couple's health, according to psychosocial epidemiologist Mara Julius. In fact, she has shown that they will live longer if they let it out![30] For shiftworking families, we must add one thing: try to schedule it outside of DZ's!![31]

### *Sharing Schedules and Time with Significant Others*

Shift schedules aren't solely for the shiftworker. Marking the shift schedule with your sleep schedule and publishing it can improve your social life! All involved will be able to plan better, watch for Dueling Zones, and develop Intelligent Time-Off Events.

### *Intelligent Time-Off Events*

Twelve-hour work schedules allow blocking of time into intelligent, high quality events and achievements. Schedules with shorter breaks don't easily lend themselves well to functional time-off events. For instance: trips, outings, workshops, family events, workshop projects, do-it-yourself projects, research projects, readings, craft shows, plays, festivals, community happenings. All of these should be tried by extra-achieving 24-hour families! These events are creative and involve family members. Shiftworkers with shorter blocks of time off may not have enough time to get involved in these extended types of activities. Still on rotating 8-hour shifts, 7 days in a row, they often find themselves isolated and bored, leading them to destructive or coping activities i.e., over-partying with other shiftworkers, extra-marital affairs, perhaps even drug use. Not the best mental or physical states for people in technically-empowered work environments.

### *Planning Custom Holidays*

One fun-loving family group swore that they held special "GollieDays" "Gollie, I didn't know that a custom holiday could be so much fun." Customizing slow-paced off-time into a special holiday can give everyone a fresh charge, especially the kids.

A 24-hour person, after working three weeks of straight twelves, can easily forget that a special day is in order. This special day should be planned before working the rigorous overtime or maybe planned after a couple of days of rest. The final exhausting day is not the best time to plan the great American GollieDay!

***Required Sleep For Both***

The entire family must realize that the amount of sleep needed by the shiftworker varies with individuals and is greatly determined by his fitness, exercise habits, and mental condition.

There is no way to rationalize out of this. No family member should pressure another into fitting into their personal sleep patterns and requirements. At the same time, the shiftworker may be getting adequate sleep, but the spouse becomes extremely fatigued by trying to keep up with the shiftworker's schedule i.e., staying awake late at night until the shiftworker gets home, getting up in the middle of the night to pack a lunch. Demanding that the spouse be awake when the shiftworker gets home is too much to ask. If the spouse has children or works outside of the home, this lack of sleep could have serious consequences on his/her attitude, alertness and responsiveness to the shiftworker, not to mention the safety issues of sleepy driving.

***Do You Have a Jungle in There??***

Getting control of the sleeping environment can be a challenge for many family members. Here are a couple of the milder tales that Peggy tells in her training sessions.

The Vacuum Cleaning Spouse — When feeling hurt or angry, or disgruntled by lack of sleep, this spouse would run a loud vacuum cleaner along the shiftworker's bedroom door while he was sleeping. This would make door banging, sucking noises that would wake the dead, giving the cleaning spouse a mild sensation of successful retribution.

The Daytime Calling Spouse — This spouse would call the graveyard-working shiftworker regularly during her daytime lunch break! The shiftworker finally broke this by surprising her several times at 2 a.m. with a "what are you doing" phone call!

## *Electronics as Off-Time Performance Tools*

Edison's light bulb isn't the only invention that can facilitate the 24-hour style of shiftworkers. No shiftworker should become isolated from the tremendous potential and growth of electronics! In fact, getting involved with electronics is a good way for the family to enjoy activities together.

### *Video Tape Recorders*

Smart 24-hour people use VCRs to record football games, news programs, movies and other programs that they would miss. Many swap out recording duties with a fellow employees on the opposite schedule! Also, video rentals can fill anyone's entertainment bill these days.

### *Answering Machines*

Answering machines can serve their real purpose of answering for the shiftworker. The outgoing message can be set to give a second emergency number or time to return the call.

### *Fax Machines and Modems*

A fax machine can serve as another method of receiving and holding messages, bulletins and planned events. These machines are now within the financial range of most shiftworkers. They may be a real necessity for those with an "off-time business" or major hobby. Most computers come equipped with built-in fax modems. A few progressive couples have found a way to communicate with each other throughout his or her shift by e-mail and faxes.

### *Pocket Pagers*

Pagers are economical enough to be used by most shiftworkers in critical jobs. And these may now be sponsored personally by the individual! By using a pager, a spouse can leave digital messages for the shiftworker. Some without alpha capability have developed numeric codes meaning: pick up child #1, #2 or #3; location #1, #2, #3; etc., and 911, call home now.

### *Cellular Phones*

Many employees can now afford cellular phones to help maintain alertness during commuting. These can also add an hour's value to each work day.

### *Internet and Computer Bulletin Boards*

The Internet and electronic bulletin boards were designed for people on different schedules to post and receive messages across the country! That design fits 24-hour performers perfectly. Most communities have at least one free bulletin board. You'll be surprised at what the personal computer and a modem can do for the shiftworker.

We'll be seeing better and more affordable access nodes to the Internet. Many companies have implemented their own intranets, a series of lan-to-lan connections within a corporate structure. The intranet has a strong fire wall, keeping external non-employees out. A company intranet is a good source of job openings, want ads, company news items, and even training. The shiftworker can dial in and download company courses to improve technical skills or development skills to improve personal communications skills. Usually within the company intranet system, you can post e-mail to other workers. Peggy now leaves e-mail for her children while she works. When the children get home from school, they pick up their friends AND Mom's e-mail at the same time!

There are many companies that not only provide Internet access, but the latest news services and forums where people discuss hobbies, work, safety and other issues. America On-Line and CompuServe are popular, although growing pains are expected as more people get into the electronic information age and decide to sign on ... all at the same time!

Another alternative to "pay for access" Web browsing, is local bulletin board systems. These are set up by computer enthusiasts and offer a means of communication, games, free or low-cost software. You can usually find the community BBS listings in your local newspaper. Give them a try! They're open 24-hours a day and they're FREE!

## SUMMARY OF F.A.S.T. TRACKING

These techniques, large and small, optimize 24-hour employee safety and performance. Certainly, we all wear different sizes of shoes and need to emphasize certain F.A.S.T. Tracking techniques.[32]

We suggest that you hold several short sessions on techniques from each of the F.A.S.T. Tracking performance areas. Or assign the job to your top 24-hour employees. It's really easy for shiftworkers to fall behind the best current techniques. As we have seen, this can quickly lead to burnout and the Double H Pattern.

You'll get good ah-ha's from a few individuals in every session. And they just may run out and buy a pocket paging service ... or join a fitness club ... or cut down fat in their diet ... or insulate their bedroom. Such techniques implemented as efforts will have undeniable performance results for your shiftworkers — and for your company! Each effort is a Win-Win situation!

Unfortunately, a random hodge-podge approach to implementing the efforts will not work to the fullest! Your employees must get into a definite Performance Tracking and Improvement approach. *This simple — but powerful method which programs the techniques as efforts quickly brings improvements.*

## Chapter Five Endnotes

[1]Many of us heard those words radioed from the moon on July 20th, 1969 to Houston by astronaut Neil Armstrong, "The Eagle has landed."

[2]F.A.S.T. Tracking, Peggy Westfall, 1991.

[3]All sleep researchers that we know believe that alcohol taken near bedtime interferes with the quality of that night's sleep. It also causes *bathroom wake-ups*.

[4]Judith Wurtman and her associates at MIT have demonstrated that eating protein (with low fat) leads to alertness while eating carbohydrates (sugars and starches) leads to calmness! This is opposite of what many had previously thought. In fact *it supports our shiftworkers* who insist that they carry something in their lunchbox besides *rabbit food.* Wurtman also identifies many foods that are neutral and do not affect alertness.

[5]See: Good news for sedentary bodies, *The Houston Post*, Reuter News Service, May 14, 1992.

[6]Also, economic experts are saying that each American employee is paying $3.00 per work hour for the total health system!

[7]From: Treasure Island, 1888, Chapter One by Robert Louis Stevenson. (The Dead Man's Chest was an island name, first used by Charles Kingsley in *At Last*, written in 1870.)

[8]One report of this is included in: *The Infinite Voyage — The Living Clock*, a single video production on PBS narrated by Richard Crenna, written by Gail Willumsen, edited by John Soh.

[9]Some sleep researchers report that *the quality disappearing in sleep-deprived personnel is not major task abilities, but the simple will — the willingness to perform routine procedures.*

[10]Some chemical plants and refineries are installing treadmills in the control rooms. Both inside and outside operators can use them to exercise and increase body temperature.

Other plants are trying computerized alertness tests. James C. Miller mentions that this area of testing is being developed further for truck drivers. It is now called "fitness-for-duty" testing and DOT is supporting it heavily.

[11]Often, a humorous Fumbling or Attention Slip by one team member will get everyone's attention and wake them up before a serious error occurs!

[12]From talks, classes, and surveys with 24-hour employees, 1969 to present. A major survey was developed and conducted with Gene Lowe of the Dow Chemical Company during 1990-1991.

[13]See: Why drivers fall asleep, and how to avoid becoming a traffic statistic. Jane E. Brody, New York Times, July 5, 1990.

[14]From: *The Sleep Better, Live Better Guide*, compliments of the Better Sleep Council, P.O. Box 13, Washington, D.C., 20044.

[15]From: Why drivers fall asleep, and how to avoid becoming a traffic statistic by Jane E. Brody, *New York Times*, July 5, 1990.

[16]Dr. Ann Romaker is director of St. Luke's Sleep Disorders Lab. From: Kansas City Lab Treats Sleep Disorders, *Houston Post*, April 12, 1992.

[17]The other five critical factors are: not smoking, limited consumption of alcohol, regular exercise, regular meals, and maintaining proper weight. See: *Wide Awake at 3:00 A.M. — by Choice or by Chance?* by Richard M. Coleman, W. H. Freeman and Company, New York, 1986.

[18]See: *How to Make Shiftwork Safe and Productive* by Timothy H. Monk, Ph.D., published by American Society of Safety Engineers, 1800 East Oakton Street, Des Plaines, IL. 1988.

[19]See: *Crew Scheduling Constraints, Appendix K of the Space Shuttle Crew Procedures Management Plan*, Mission Operations Directorate, Operations Division, Basic, Revision A, Jan. 1990.

[20]This isn't always done with success.

[21]It is important to remember that the sensitivity to daytime environmental noises, light, temperature, and disruptions is greatly increased when the internal clock is disrupted by a maladaptive schedule. From: *Biological Rhythms and Rotating Shiftwork — Some Considerations for Air Traffic Controllers and Managers* by Carlton E. Melton, Ph.D. and Robert S. Bartanowicz, Ed.D., Air Traffic Plans and Requirements Service, U.S. Dept. of Transportation, Office of Aviation Medicine, FAA, 1986.

[22]The first sleep disorder center opened at Stanford University in 1969. From: *Wide Awake at 3:00 A.M. — by Choice or by Chance?* by Richard M. Coleman, W. H. Freeman and Company, New York, 1986.

[23] *The Sleep Better, Live Better Guide*, The Better Sleep Council, P.O. Box 13, Washington, D.C., 20044.

[24]From: Stressed-out Americans try to balance home and work lives by Catherine Dressler, Associated Press, March 22, 1992.

[25]From: Work shifts of full-time duel-earner couples: Patterns and contrasts by sex of spouse by Harriet Presser, *Demography*, 24, 1987.

[26]The three-year study by White and Keith used a national panel of 1,668 married men and women who were interviewed in 1980 and again in 1983.

The effects of shiftwork were assessed on six measures of marital quality: marital happiness, interaction, disagreements, general problems, sexual problems, child-related problems, and probability of divorce. "Every indicator of marital quality (was) significantly and negatively affected by shiftwork in at least one analysis." *Shiftwork was found to increase the probability of divorce from 7% to 11% over the three-year period.* From: The Effect of Shiftwork on the Quality and Stability of Marital Relations by Lynn White and Bruce Keith, The University of Nebraska-Lincoln, *Journal of Marriage and the Family*, May 1990.

[27]From: The Second Shift: Employed Women are Putting In Another Day of Work at Home by Arlie Hochschild, Tish Sommers, Laurie Shields, John P. Robinson, and Mary Stewart Van Leeuwen in *The Utne Reader*, April 30, 1990.

[28]We are tempted to ask why Supervisors do not tack onto the shiftworking mother's time card an additional 12 hours per week. This would give a realistic view of her work load and possible fatigue level! Forty-two would give her 54, the minimum! Fifty would give her 62! Incredible.

[29]From: *Wide Awake at 3:00 A.M. — by Choice or by Chance?* by Richard M. Coleman, W. H. Freeman and Company, New York, 1986.

[30]In a study of 192 married couples, researchers at the University of Michigan at Ann Arbor found that when both husband and wife suppressed anger, the mortality rate for women from heart attacks ran at 11 percent. The rate dropped to 7 percent when only the husband expressed anger. When both "let it out," the female death rate dropped to zero. Also anger-suppressing husbands were about twice as likely to die of heart attacks as husbands who let out their emotions.

Mara Julius says that sulkers, holding in their anger, have the highest risk of death. She says this is because these people also tend to have such unhealthy habits as smoking or alcohol abuse. From: Developments to Watch, *Business Week*, April 27, 1992.

[31]Peggy and Glenn think they have become proficient at identifying "nuked employees." These are individuals who have experienced an unscheduled family blow-up a few hours before coming to work. Our *nuked individuals* give poor responses, appearing mentally fatigued and depressed. Looks like another Set-Up for impaired performance and human error!

[32]Use the summary list of F.A.S.T. Tracking Items in the Appendix.

# CHAPTER SIX

## Promoting and Using Performance Tracking and Improvement

### OBJECTIVES OF SECTION SIX

After completing this section, you will be able to begin a valuable Performance Tracking and Improvement Program. Much of the program can focus on techniques covered in the previous section.

In this section, we explain the most powerful technology for 24-hour employees — that of optimizing performance no matter what the schedule! Ideally, it is good to assess current scheduling and look at ways of improvement, but in many facilities, the schedule is temporarily etched in concrete. Is the situation hopeless? Not necessarily so. There is much we can do for the people dedicated to these schedules.

We will explain the simplicity of "F.A.S.T." performance tracking and improvement and give you ideas on how to promote it among your employees. The system's science relies upon tracking techniques as "efforts" in each of the four F.A.S.T. Tracking performance areas. Setting up an individual program is very easy ... just get ready to view a few of the shiftwork techniques we've discussed as efforts to achieve targeted results and you're prepared.

## THE TRACKING SYSTEM

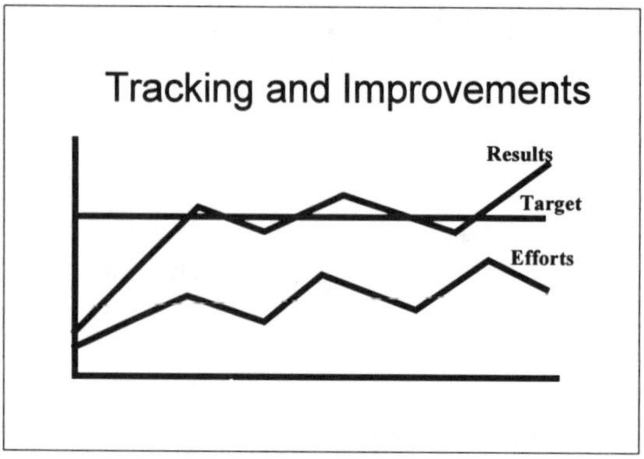


For a complete operating behavioral system, we divide the performance tracking system into three recording categories: Targets, Efforts, and Results. The plotting of these three items on Tracking Charts is a simple but powerful method of effecting personal (or group) improvement.

Performers using the system quickly get "training results" from three behavioral effects: the attention effect,[1] the measurement effect, and the applied effort effect. Add these up, and you've got a synergized performance powerhouse!

**Performance Powerhouse of Effects**

- Attention Effect
  - Performer's attention, boss's attention
- Measurement Effect
  - Measuring, tracking targeted efforts, aiming at target results
- Applied Effort Effect
  - Actually doing it!

Performance Tracking and Improvement is applied behavioral science at its very best: simple, trackable, real-world oriented, and directed at the individual's performance improvement. And best of all, it is driven by the employees desire to improve. It seems like such a long way, getting employees to this stage. But, once the employee can recognize himself as we have discussed so far, he is usually looking for ways to feel better, to enjoy his time off, to enhance his relationships, and most importantly, to improve his performance at work.

Instead of diving deeply into behavioral science and the powerhouse effects (which could take a while), we're going to introduce you to Bill Whitley who experienced them directly. Bill got into F.A.S.T. Tracking so quickly and thoroughly that he makes an excellent example.

**F.A.S.T. Tracking Bill Whitley**

Bill Whitley[2] is a 34-year-old Senior Plant Operator who used F.A.S.T. Tracking to reach a personal and job performance level he had previously thought impossible. It all started after Bill had attended a F.A.S.T. Tracking. We discussed fitness and how it can affect not only his wellness, but the quality of his sleep, his energy levels, the quality of his time off, and other areas. He had noticed how lively the eight new hires looked and how much energy they put into the session. He convinced himself to give F.A.S.T. Tracking a try since several of the others said they would.

The good thing about F.A.S.T. Tracking, he thought, is that it was HIS program. No one would monitor his efforts, but they might notice the results! Hey, it might be fun! He went by the drugstore and bought himself an inexpensive planner to keep some personal notes.

Bill first developed a snapshot of himself, we call a **Current Track**. Over two months, he observed and tracked a few important performance indicators that would snapshot his current condition.[3] He said that remembering to debrief himself and track those items each day was " ... about the most difficult thing I've ever done and several folks thought that I was crazy!" However, he enlisted fellow shiftworkers and family members who good-heartedly jogged his memory often.

**The Current Track**

Although he appeared to be in excellent health, he found that his weight had crept to 170 pounds. That was kind of high for his small build and height. He was continuing to fill his 35-year-old body with his beloved fried foods (in spite of what experts were saying on television), and his fat intake was probably too high; his golf score had plateaued at 95 for a year; he was experiencing heavy fatigue and jet lag about six days a month on his 8-hour backward rotating schedule; and finally, he was averaging only 2 dates per month, which seemed a little low to him as a single guy. Now, that was a quite a snapshot!

**Bill's Current Track:**

- Weight 170
- Eating Fried Foods
- Golf Score: 100
- Fatigue/Jetlag: 6 days/month
- Dates: 2/month

Bill's first experiment was to cut out all fried foods. It seemed simple, and soon he noticed it wasn't too difficult. He would just prepare his food differently or order the right kind of menu food. He then became interested in moving into a more serious performance improvement program where he would set **Targets** and then regularly debrief his efforts and see results.

Since Bill felt that he was a career shiftworker, his **Target Areas** became the improvements that he would attempt in fitness and sleep. He decided that he would focus directly on his fitness. He had been told that this would have the **Spin-Off Result** of improving his sleep.

He conferred with the Plant Nurse who encouraged him and gave a couple of suggestions. She didn't claim to be a fitness expert but mentioned that he was getting a little heavy considering his height and body type.

After he saw the nurse's enthusiastic response to the Tracking Charts which could give him feedback on his efforts, Bill jumped wholeheartedly into the program. After all, he needed to lose a few pounds, and this didn't cost anything. He told a few folks that he was getting into the program, and he jokingly disclosed part of his goals with them: reducing weight, toning some muscles, improving his sleep and getting a few more dates!

It was time to write the **Tracking Plan,** which was a brief statement of his targets and the efforts that he would make. Bill placed the plan in his planner where he would debrief and track himself daily.

He wrote the following targets to hit within three months:

**Bill's Targets:**

- Weight: 160
- Jog: 12 miles/wk
- Stop Fried Foods
- Calories: 2500/day
- Better Sleep
- Golf Score: 90
- Fatigue/Jetlag: 3 days/Mo
- Dates Per Month: 4

Bill then plotted his targets while sifting through the stack of blank Tracking Charts that had been handed out during the F.A.S.T. Tracking session. He found pre-made charts to help measure his efforts toward reaching each target. "Nothing to it! Weigh in per week, miles jogged per week, grams of fat eaten per day, sleep per day!" He later showed two of the charts to the five employees on his shift (weight and jogging). He figured that was plenty to divulge and he would receive plenty of ribbing for doing even that.

As Bill began to track his efforts every day on the charts, he began to see progress "across the board." The performance powerhouse of attention, measurement, and efforts turned out to be a better motivator than anything — including Bill's family, peers, boss, and doctor.

Bill divided this chart into weeks so that he could see his weekly jogging totals more easily. The weather was pretty warm in September so he also set upper limits to ensure that he wouldn't overdo it at this point. The first month of jogging looked pretty good to him!

Listening to friends' advice, he tried the calorie counting route. It was easy to set up on a chart but was a pain to do everyday. With the aid of a calorie calculator, he could keep a good average and stay below the upper daily limit. He decided next month to focus his efforts on other areas less tedious and more motivating. Counting daily fat grams proved to be much easier than calorie counting.

Bill felt that his weight was perhaps the most important target to watch. Amazingly, the one-month trend headed in the right direction! His efforts in jogging and calorie counting were showing a payoff at this early stage.

Bill tracked his sleep and schedule for comparison. His sleep continued to have that erratic downturn during graveyards. It was too early to tell if improved fitness would help here. Bill felt that he was sleeping better, but he couldn't prove it yet.

Bill's golf game was his passion. He hoped to become a competitive senior-class player in the future. After seeing the first full month of efforts and results, he planned to develop a program to increase efforts during months of good weather — especially during vacation time.

Certainly, Bill's tracking and performance improvement began with more ambitious targets than those of his fellow plant operators. Bill had quite a time tracking and debriefing himself on the items. Fortunately, this plan resulted in achievement of his (very practical) targets in just three months. He had planned to see a fitness specialist or sports physician if things were progressing too slowly. He would have asked the specialist to help him plan a better set of efforts. But, that visit turned out unnecessary!

After that initial three months, Bill learned from each positive attempt he made and continued to work at improvement. During that first three months, the three components that Bill plotted on each Tracking Chart were:

**Targets**: specific results Bill attempted in one of the four performance areas: fitness, alertness, sleep, or time off. Targets may be private or team based. Team-based targets are shared on the job. This is strictly up to the individual. Targets are usually entered as straight dotted lines on the tracking charts.

**Efforts**: simple measurable efforts that Bill made each day, week, month, etc. to ultimately achieve a result. Efforts are entered daily.

**Results**: are what Bill achieved from the efforts. A result is the final entry or final average on the chart. Hopefully, it matches the target level!

**Creating F.A.S.T. Tracking Targets**

Good targets, like good training objectives, are measurable, therefore. Targets should be formulated in numbers-over-time such as: hours of sleep per day, number of pounds per month, number of calories per day, number of aerobics points per week, or sports points per game.

*Examples of Targets*

The following illustrations give examples of measurable-trackable targets that other individuals have chosen to aim at. Show these examples to your employees. You'll be surprised at how few takers you get immediately and how many you will get in the long run after they see a couple of employees get going!

*Performance Improvement Tools*

As you can see, the tools are simple: F.A.S.T. Tracking Targets and F.A.S.T. Tracking Indicators (Efforts and Results).

The language is also simple and user-friendly. For instance, we have found that the term "performance improvement program" carries much more appeal than "behavioral change."[4] The real behavioral science-based approach relies upon scientific tracking of targets, efforts and results, not upon intimidating buzz words. **We can keep the program powerful and useable by keeping the language simple for everyone!**

*Tracking Indicators (FTIs = FTEs + FTRs)*

There are two types of F.A.S.T. Tracking Indicators (FTIs): efforts (FTEs) and results (FTRs). In behavioral science terms, efforts and results are spoken of as "causes" and "effects."[5]

*FTEs or F.A.S.T. Tracking Efforts*

FTEs are indicators of the participant's efforts (measured causes) in fitness, alertness, sleep, and time off. These efforts are often taken from popular accepted techniques. Typical efforts are daily aerobics points, biking miles, calorie control, hours of strenuous yard work, and hours of quality time with family/friends. We know that most of these efforts lead statistically to results!

*FTRs or F.A.S.T. Tracking Results*

Over a period of time and effort, the recorded efforts (causes) will lead to recordable results (effects).

The individual taking active control and recording his/her aerobics points, fat intake, and total calories in the Fitness Area will achieve results such as improvements in pulse rate, subjective ratings of higher energy level, better sleep, and better ratings of alertness while on the job.

**Who determines the targets, the efforts, and the scorekeeper?**

We promote employee determination of tracking and improvement. This is vital to helping the employee increase personal control over performance. This doesn't mean that you should not move into the coaching role with some employees.[6] Certainly, employees often allow experts and peers to influence their efforts. However, your employees will work for ultimate performance only when they actually own, control, and get rewards for that performance![7]

Continuous performance improvement is an opportunity where you must take the role of leader and supporter — not director, authoritarian, "expert" or even floor supervisor.

**Team-Based Tracking**

Team-based Tracking is where the entire team takes on improvement targets in Fitness, Alertness, Sleep, and Time off. The climate of your organization, determines whether you should (and can) focus on Team-based Tracking. (If you're experiencing Darkside Morale, it may be a little early to promote this.) Firemen, police, and other emergency response personnel are often open to competitive group physical programs that will obviously increase their job effectiveness. On the other hand, established groups in some refineries and plants will tell you to take a hike ... that their health and time off efforts are their business! However, team-based efforts will lead to team-sized results! If a little competition can be introduced and tolerated, the amount of positive team results will amaze you.[8]

**EMERGENCY RESPONSE TEAM EXAMPLE**

For instance, soon after beginning a team program, an Emergency Response Team can begin recording F.A.S.T. Tracking improvements as in the following examples:

Fitness - average ratings of the group's physical ability, speed ratings from team emergency drills (tracked from month to month).

Alertness - ratings of average team member alertness, as measured by various electronic instruments; team scores on electronic games/tests during Danger Zone play.

Time off — sports competition scores and hobby competition. Competitive run-offs can be highlighted at company picnics and barbecues.

## Targeting, Trying and Tracking!

### Targeting and Tracking Effect

It's amazing how often improvements can be made by the simple technique of targeting, trying, and tracking! Certainly, this combination takes advantage of the attention and measurement effects that we mentioned earlier. Tracking often leads to the problem being mysteriously solved ... without exerting clearly known efforts! It also leads to a fresh look at self.[9]

### Trying — Using the Placebo Effect!

In our team-based safety programs, we've come to realize that we often obtain what many researchers would term a "placebo effect." At first we were concerned that this effect was undesirable and would throw us off track. However, we have witnessed many positive unexpected results and now feel that the effect is highly desirable in the typical human performance area. The increased attention and efforts in a targeted area leads to a temporary breaking up of old behavior patterns.[10] Any of a number of efforts in this area will cause improvement, if one can simply break up the autopilot behaviors for a few days.

### CRISIS LEADS TO CHANGE

Behavioral science often forgets its debt to the simple pleasure-pain concept.[11] We can really look at all behavior change as being motivated by those reward and punishment twins ... pleasure and pain.[12] We try new efforts in order to achieve pleasure and avoid pain. Peggy talks about how group 12-step programs teach us that "hitting bottom" (feeling intensive pain) is necessary before a well-rooted behavioral pattern can be broken.

Much pleasure and pain interpretation is due to attitude and belief.

## ATTITUDES AND BELIEFS

Unfortunately, many influential behavioral scientists have tended to discount attitudes and beliefs and what goes on inside the individual since such variables are not directly observable. **These non-observable critters are considered difficult to attack head-on via behavioral technology for some strange reason we have never figured out.**

Nuclear science, astronomers, physicists, have worked for years with items thousands of times smaller, larger, and more distant than human variables! Think about it! In spite of this non-visibility, those quarks, hidden galaxies, genes, and magnetic fields have been identified, studied, and cultivated.

## INVISIBLE BUT REAL!

- Quarks
- Galaxies
- Genes
- Magnetic Fields
- Attitudes
- Beliefs

Could it be there's still a little bucolic backwardness in behavioral academia? Before attempting any behavioral change program, supervisors, psychologists and trainers should prepare for the following responses which indicate attitudes and beliefs that will stop 'em dead in their tracks.

## PERFORMANCE IMPROVEMENT (BEHAVIORAL CHANGE) BLOCKERS

**I can't.**

"Although I know I should quit, I can't," is a response heard everyday from smokers, drinkers, and eaters.

**Yes, but!**

"I should try it (a new effort) but," will smash program efforts in a heartbeat.

**They'll think ...**

"They'll think I'm crazy" is heard from employees having such a high regard for peer pressure that they will try nothing.

## FIRST STEP: CONVINCE 'EM!

The lesson for corporate managers and trainers is simple: Never discount the importance of attitudes and beliefs, no matter how "behavioral" we attempt to become! Our employees are attitude and belief driven, nothing more — nothing less. Your first step in developing a shiftwork program is to "convince 'em!"

## Beginning your own Shiftwork Safety and Performance Program

### Warming Up with Communications

To begin your own Performance Tracking and Improvement Program, run some warm-up sessions by using the overheads from this book. You may want to first get your top management enthusiastic about the shiftwork improvement issue. Next, you'll want to tackle your 24-hour people in two or three communications sessions. Call us if you have tough problems or need live assistance.

### Developing a 24-Hour Committee

One item that works superbly to catalyze our F.A.S.T. Tracking programs is ... the F.A.S.T. Tracking Committee, a committee spanning different functions and utilizing employee involvement.[13]

A committee consisting of "Reps" from different functions and plants can put tremendous front-end energy into a new shiftworker's performance tracking and improvement program.

Look at this typical enthusiastic committee: a Safety Manager, the Medical Director, three Operators (from each production plant), and the Operator Training Coordinator. Such a committee can move things quickly. Here is the action list developed by one committee at their very first meeting:

ACTIONS TO PERFORM THIS MONTH

- Send bulletin board announcements to all departments.
- Solicit coordinators in each 24-hour department. Have dept. coordinators then enlist a coordinator for each shift.
- Enlist a representative from the scheduling department.
- Schedule Kick-Off Presentations/Games/Prizes for each shift/each plant.

ACTIONS FOR NEXT MONTH

- Plan an immediate Follow-Through Session on a Fitness subject. (Medical Director will handle this first one.)
- Appoint Follow-Through Session Sub-Committee. This committee will plan more bimonthly sessions plus monthly bulletin board and night shift Safety Huddle emphases.
- Get with Company Picnic Committee to work the F.A.S.T. Tracking theme into the spring picnic in some way to include spouses and significant others.
- Discuss creating a F.A.S.T. Tracking Contest Sub-Committee to promote tracking of shiftworker aerobics, sports, and other off-hour fitness efforts.

## HANDLING LEARNER READINESS IN 24-HOUR PEOPLE!

One of our previous program research areas has been a trainer's concept that we term "learner readiness." We've watched learner readiness in F.A.S.T. Tracking participants very closely, since we want to take every advantage of this phenomenon.

Learner readiness (LR) shows great variability among shiftworkers. As you would guess, it runs tremendously high in brand new shiftworkers (and their families). The LR Curve then follows either the Burnout Cycle or the Achievement Cycle. With three-year workers, many have given up completely on optimizing their own performance. Skepticism has set in and learner readiness has disappeared. They've moved into the Double H Pattern.

The first three years of 24-hour work sets the learning stage for the career shiftworker. **During this first 36 months, the shiftworker invents a good personal performance system or falls into the hopelessness and helplessness trap.** We are convinced that without a special training push, the majority will lose out to this pattern of personal and family hopelessness.

The most learner-ready workers over three years are those that have developed some sort of personal performance system that is working for them. It seems that achieving noticeable previous successes keeps them open to learning about F.A.S.T. Tracking techniques! We're convinced that these folks will end up achieving even more and will become the lead role models for new F.A.S.T. Tracking employees.

Peggy says that getting them ready to learn in a brief classroom session feels like preparing them for a jogging session. If they are learner ready, e.g., successful or preparing for a new job, they will be ready and willing to jog. However, if they don't feel that they should be in training, then it's like dragging them onto the starting line!

## Learning Through Optimism and Encouragement

High learning on the job occurs when the boss, family,[14] and peers encourage and reinforce performance. Success becomes imminent in this achievement model.

**The best behaviors — those most desirable efforts and results — occur only after repeated, multiple training sessions and reinforcements.** In other words, all must **keep at it** and not expect one-shot programs or fads to work a miracle!

## Getting Them Involved!

Your success depends upon getting your employees' involvement in their own 24-hour safety and performance program. Employees must at some point jump-start themselves and take control of their own safety and performance. As you probably have learned through experience, there just isn't enough of you, as an individual supervisor, to do it for them! As our good friend Charlie Kilsby says, "There's only one of you ... there's six, twelve, twenty or more of them!"

## STRATEGY #1: INCREASE YOUR EMPLOYEES' INVOLVEMENT

In fact, you will find that 24-hour employees, once set in the positive direction, will shoulder a good program themselves. It is our basic hypothesis that SHIFTWORKERS ARE REALLY EXTRA-ACHIEVERS. This can work for the program if moved in a positive direction, or against them, the program, and their company if pushed to the darker side.

### Using the Existing Teamwork of 24-Hour Workers

The nature of 24-hour work makes it an area that requires both self control AND positive teamwork by your employees. We believe in the power of "team-based behavior change." The entire team can get into performance improvement if you manage it the right way.

**One thing for sure ... We'll bet that your night workers are already operating as a team.** Their teamwork, if operated against you (read: as negative peer pressure), will nullify any program or management practice quickly. If operated in favor of a positive safety and performance program, it will take you to home base!

## STRATEGY #2: USE EXISTING TEAMWORK TO PUSH THE PROGRAM

## SUMMARY — LEARNING THE ESSENTIALS FROM NASA

When we first walked through the halls of NASA, we spotted posters and bulletins that showed beyond a shadow of a doubt, that the entire U.S. space team WAS A TEAM and was committed to performing their very best — committed to reaching a common goal.[15] **Likewise, your employees MUST be energized and motivated before any program will take hold.** The NASA team was certainly motivated. What could be more inspiring than the prospects of orbiting the earth, landing on the moon, and developing the space shuttle? Since the time of that first visit to NASA, our programs carry more motivational pizzazz ... and why not?

As we talked with NASA personnel responsible for the ultimate shiftworkers (the shuttle astronauts), we realized something that should have been in our trainer's minds since the beginning — **research is good - but practical application is where it's really at!** And there it is ... Our trainees and fellow managers can listen to the laboratory research about circadian rhythms, jet lag, and sleep, but actual proven techniques that work for them is what will stick in a program.

Since those conversations with NASA, we've made it a point to emphasize the practical application, and take the research for what it is — knowledge that needs testing under field conditions. You'll see very few experts running our programs that haven't actually worked the job!

Finally, we reviewed the process that NASA used to improve their crew's flight behavior ... They would try something new during a flight and then debrief the crew's performance immediately upon re-entry landing. From this knowledge gained, they would modify certain behaviors for the next trip. Simple! In manufacturing processes, we have been calling this "continuous improvement!" In this manual though, we're not talking about manufacturing processes. We're talking about improving human performance.

**To manage performance improvement on a group or personal level, you absolutely must track efforts and results, then set your targets again.** It's as simple as that! It's done with astronaut teams; it's done with football teams; and we can do it with our own performers in those critical jobs.

Certainly other managers and trainers have something to learn from NASA and from behavioral science. Look at the many commercial programs on the market that have no targets, no follow-up, nothing to track, nothing visible to reinforce or improve over time. Unfortunately, the ones that take the time to include these items are usually so dull, boring, and/or clinical that they waste your time and the time of your employees!

**So have a go at it — try your best!** Begin with a small or large program and make an important impact on a critical job! While you're building it, make the program as lively as possible ... with excellent sponsors, instructors, bells, whistles, and experts. And let us hear how you did ... and how they are doing — those most important 24-hour performers!

## CHAPTER SIX ENDNOTES

[1]In management training, the attention effect has often been billed as the *Hawthorn Effect*, cited from studies at Western Electric in the earlier part of the century. Until well after those experiments, "employee involvement" and simple attention to employees was not rated highly by organizational *experts*. The *Placebo Effect* is a related effect that often works positively but is not utilized well by the current behavioral sciences. (Many will disagree with this statement!)

[2]Bill is a composite from several of our 24-hour performers!

[3]Behavioral scientists term this "base-lining."

[4]Although in some organizations it may sound derogatory (for lower-rated performers)!

[5]In performance improvement language, "efforts" sound more like something manageable by the individual human where "causes" sound to us like something very big, mysterious, and impenetrable. "Effects" sound like the outcome of a lab experiment or maybe the outcome of a disease to many of your employees!

[6]See team-based tracking immediately following.

[7]Someone once said, "This is like behavioral free enterprise!"

[8]See *Secrets of Team-Based Behavior Change Programs* in the Appendix.

[9]Glenn was first exposed to this behavioral area at the young age of 19 in the U.S. Marine Corps. Using a mail order "self-improvement" program, he learned to baseline and chart his time management, to touch-type, to take several high-school courses and to push the general limit all within a year. At the end of this 12 months, he had typed a 300-page manual, had written several training articles — and had won the position of *Training NCO*, progressing two ranks. Not bad progress for a ninth grade education at the time!

[10]Our safety programs label those actions that become automatic and performed without thinking "autopilot behavior." Attention to an autopilot behavior temporarily breaks up the autopilot condition-response and can cause a surprise improvement. For instance, taking a close look at a tennis player's serve causes the player to temporarily pay attention as he/she is performing the serve. (We also hear of negative effects from too much attention in the sports area! Meaning: don't mess with too good a thing!)

[11]Various writings of Sigmund Freud early in the century.

[12]*Awaken the Giant Within* by Anthony Robbins, Simon and Schuster, NY, 1991. (A well-written, best seller popular with trainers!). Pain or pleasure! Everything you and I do, we do either out of our need to avoid pain or our desire to gain pleasure."

[13]Some companies are using existing safety committees which already include the necessary 24-hour employees and department representatives. Assigning the additional shiftwork emphasis to them may get them very excited and energized.

[14]We label the family and close friends as the *primary support system.*

[15]Dr. Charles Garfield has written some wonderful materials on the subject of teamwork and performance. He learned about the characteristics of top performing teams during his stint at NASA.

# APPENDIX ONE

## Surveys and Checklists

# Supervisor's Checklist for 24-Hour People

[ ] Employees can discuss Shiftwork Set-Ups and Slips.

[ ] Employees can predict effects of moving the sleep schedule too rapidly and can recall the maximum desirable movement.

[ ] Employees have analyzed their schedule rotations for heavy jet lag areas.

[ ] Employees understand necessity of developing team alertness devices. Several report using one or more methods.

[ ] Employees can recall the effects of fitness upon 24-hour alertness and performance.

[ ] Employees can describe the effects of food, light, and sleep upon Danger Zone alertness and performance.

[ ] Employees can describe the main sleep environment items.

[ ] Employees can discuss the effects of time off and the "primary support system" upon performance.

[ ] Employees can describe the concept of Performance Tracking and Improvement.

# F.A.S.T. Tracking Items

**Fitness**

1. Exercised one hour today!
2. Exercised! Within one hour of waking.
3. Ate lighter meals during evenings and midnights! Right before and during shift.
4. Progressed to mostly carbohydrates around midshift.
5. Minimized sleeping pills, coffee, alcohol, tobacco.
6. Consumed no alcohol near bedtime.
7. Other: ____________________

**Alertness**

1. Changed pace often!
2. Took breaks often!
3. Refreshed myself! With light snacks.
4. Stepped outside for fresh air.
5. Handled Danger Zone:
   · Stopped and thought twice before important decisions.
   · Used checklists and procedures intensely.
6. A.Q. tested team members.
7. Other: ____________________

**Sleep**

1. Preplanned my sleep schedule.
2. Analyzed work schedule.
3. Minimized moving my sleep hours.
4. Slept my minimum hours.
5. Slept in a dark, quiet room.
6. Other: ____________________

**Sleep Environment**

1. Published my sleep/work schedules for family/friends.
2. Turned off phone and doorbell ringers.
3. Had answering machine catch calls.
4. Used "Call Forwarding" to catch calls.
5. Kept sleep room at about 68 degrees.
6. Added white noise (fan, radio, television).
7. Other: ____________________

# F.A.S.T. Tracking Items (cont'd)

**Time Off**

1. Maintained good sleep schedule during time off.
2. Told family/friends times when it was okay to awaken me.
3. Spent "quality time" with family/friends.
4. Scheduled <u>well in advance</u> to synchronize with their eating, recreation, movies, sports, social activities.
5. Recorded, later watched sports games, movies, documentaries.
6. Paid attention to sex life.
7. Other: ______________________

# Secrets of Team-Based Behavior Change Programs[1]

**HAVE AN ISSUE IMPORTANT ENOUGH TO WARRANT REAL TEAM WORK.**

In American organizations, it takes an extremely important issue to overcome our front-line team member competitiveness and individualism. The average American would rather compete any day! Issues that sometimes "get through" and promote intensive teamwork include group and individual survival such as serious economic, environmental, and physical threats. How many times do you see a group involved and 100 percent behind a new issue? If you can get 80 percent of the team members behind your program, you will excel!

**Persuade 80 percent of the team members to get involved with the program from the beginning.**

This is done through kick-off meetings and promotional announcements. Don't plan to coerce every individual to the acceptance point. Only team/peer pressure can eventually get everyone to "work the program." Expect resistance on any program that causes extra work or true behavior change.

**Speak up, not down, to them.**

This is crucial. Have you ever responded to a program that had you personally on the defensive or made you feel childish? When our safety and quality programs seem "hokey" to the most important team members, we need to look closely at the way we promote them.

**Have team members contribute program items: ideas, procedures, follow-ups, personal time.**

Brainstorming sessions that encourage input will commit your employees to the program. They must feel that they are really part of the team-based change program. It's surprising what employees can do ... if they truly feel they've helped in the change program. It may be your management goal, but it's their program.

**Expect incremental improvement, not overnight success.**

Small incremental changes and improvements will occur over a period of time. Many will not be noticed or reported unless you have good detectors! And when you do ...

**Catch 'em doin' it well!**

The best feedback to them is simply to catch them doing it well — a certain targeted TII (incremental improvement.) The TII can be observed by you or self-reported by them.

# Shiftworker's Survey

*(This questionnaire is similar to one we used prior to kicking off the F.A.S.T. Tracking training sessions.)* This questionnaire is to get your input on a completely new training program designed for shiftworkers. This session will be instructive, positive, and entertaining. It will be valuable for those employees just entering the field and those who have worked shiftwork for 30 years!

Shiftworkers have become a large and valuable (sometimes neglected) portion of our working population: Plant and Field Operators, Pilots, Medical Personnel, and Police Officers work around the clock.

We have observed employees who have developed positive ways of dealing with shiftwork. Using the data you provide us plus information from other studies, your training session will stress positive techniques that should be applied by people in critical 24-hour operations. Please answer any of the questions you can.

**Humorous Incidents**

List any humorous incidents you have experienced due to shiftwork. For example, SlipWalking, such as forgetting an appointment, going to work at the wrong time, etc.

____________________________________________________________

____________________________________________________________

**Safety and Health Issues — How you deal with them.**

List safety and health issues that were caused by the odd hours. (SlipWalking, weight, energy, thinking, etc.) List techniques you successfully use to deal with them.

____________________________________________________________

____________________________________________________________

**Family and Friend Relationships — How you deal with them.**

List family/social issues that develop due to your shiftwork. Tell us how you successfully deal with them (i.e., missing dinner with them, missing Saturday night events.)

______________________________________________

______________________________________________

______________________________________________

**Positive Benefits and Challenges**

How has shiftwork benefited or challenged you (hobbies, school, travel, etc.)?

______________________________________________

______________________________________________

**Your Suggestions**

List your other positive ideas for dealing with shiftwork that you feel will work for others. For example, special diet, special sleeping habits, special exercises, ways of dealing with home maintenance.

______________________________________________

______________________________________________

**Information About You**

Years you have worked shiftwork: ____________

Your age: ____________

Your name (optional): ____________

Date: ____________

# APPENDIX TWO

## Incidents and Costs[1]

The following collection includes incidents and accidents in 24-hour operations allegedly caused by individual and team error. Admittingly, this list is incomplete and actual costs and details have been summarized from media and government reports.

4/16/47 Texas City Explosion — French ship Grandcamp exploded while docked. Ammonia nitrate exploded. On the next day, more fertilizer on the High Flyer blew up. In all, 576 fatalities and 5,000 people injured.

9/29/57 Soviet Nuclear Incident at Chelyabinsk (near Kyshtym in Urals). Apparently, a tank of radioactive waste exploded, discharging 20 million curies of radiation. Probable contamination of 357 square miles, 10,000 people evacuated. Possibly several hundred fatalities. Est. 200 million rubles.

12/27/77 Farmer's Export Grain Elevator Explosion — Grain dust was ignited by a spark. There were 18 fatalities.

3/28/79 Three-Mile Island, Middletown, Pennsylvania — Nuclear reactor partial meltdown due to combination equipment error and human error.

11/1/79 Galveston Bay Oil Spill — Tanker spilled 10,700,000 gallons of oil. Two ships had collided in the Bay.

10/13/81 Polyethylene Plant Explosion — Dow Chemical, Texas suffered a major explosion that occurred during a shutdown. There were 6 fatalities.

11/19/83 Mexico City Gas Storage Explosion — Four spherical 420,000 gal. tanks were ignited from a propane truck at loading dock. Homes were allowed to be built near the facility. Thirty acres of homes destroyed and another thirty acres damaged. There were 540 fatalities, over 2,200 people injured, and over 10,000+ people left homeless.

1984 Houston Nursing Home Incident — An eighty-four year old woman was strangled by a restraint device. Human error in a medical setting — extended over three nights. $40,000,000 was awarded to the family of the woman.

2/25/84 Vila Soco Pipeline Fire — Pipeline gasoline blaze exploded and burned at over 1000°C through a Brazilian village. The wrong pipeline was opened the day preceding the fire. There were over 500 fatalities. Child casualties under the age of 5 had to be estimated since they were totally incinerated. Petrobras paid hospital costs and damages.

7/30/84 Louisiana Coast Oil Spill — Tanker Alvenus spilled 2,800,000 gallons of oil. The ship struck bottom and ruptured off the coast of Louisiana. An 85-mile oil slick drifted into Galveston Island and Port Arthur shores. Galveston beaches and Port Arthur coastlines were damaged. Foreign owner paid $1,100,000 to the State of Texas to settle.

11/19/84 San Juanico Pemex Gas Explosion — A series of LPG storage explosions in San Juanico, Mexico. Fireball flashed through a suburban area at 5:43 am. There were 503 fatalities, 4,000+ injured. Pemex was held liable by the federal attorney general. By 1986, Pemex had paid $5,000,000 in claims.

11/5/85 Mont Belvieu Plant Explosion — Warren Petroleum plant exploded. One hundred fifty workers reached safety minutes before the ignition! There were two fatalities.

1/6/86 Kerr-McGee, Gore, Oklahoma Nuclear Mishap — Cylinder of nuclear material burst. Cylinder was being improperly heated. There was one fatality and 100 people were hospitalized.

1/28/86 NASA Challenger Space Incident — Shuttle failed due to a faulty booster rocket O-ring. Sleep loss and overtime were cited as key set-ups for this accident that killed all seven astronauts. This incident significantly impacted progress of NASA programs. Morton Thiokol changed name to Morton International, Inc. due to negative publicity, and invested $600,000,000 in automobile air bags to improve image.

4/26/86 Chernobyl — Nuclear plant meltdown and radiation release. Cause included poor engineering and operator error. The accident released 50 million curies of radiation into the surrounding area. At least 250+ deaths and many more are expected due to the lasting acute and long-term effects of high level radiation. $26 billion is planned to move 200,000 additional residents. $2 billion is planned to rebury the plant.

7/8/86 Ohio Tank Car Gas Release — CSX train derailed leading to explosion and burning chemicals. Tank car of white phosphorus released gas over

Dayton suburb. Twenty-five thousand people were evacuated. Governor called for more state regulations.

10/30/87 Texas Refinery Acid Release — Hydrofluoric acid released from Marathon Petroleum refinery led to evacuation. Two hundred twenty-five people were injured and 4000 residents were evacuated.

1/2/88 Monongahela River Spill — Ashland oil tank ruptured and spilled 3,800,000 gallons of diesel fuel. Over 860,000 gallons escaped into river. At least 70 communities lost their water supply.

1988 Conrail Train Collision Thompsontown, PA — Two freight trains collided head-on killing four persons. Engineer and crew possibly sleeping or experiencing micro-sleep. Over $6,000,000 in damages.

3/24/89 Exxon Alaskan Oil Spill — A 987-foot tanker smashed into Bligh Reef and spilled 11,000,000 gallons of oil into Prince William Sound. Capt. Joseph Hazelwood left the bridge during maneuvers. He and his crew have been blamed by government officials and others. This incident was eventually labeled as a HUMAN FATIGUE incident by the NTSB investigators.

On-going: The U.S. Congress passed a bill allowing states to adopt stricter spill liability laws than the federal government requires. (Exxon later felt that a Sperry control system might have caused the incident.)

6/23/89 Houston Ship Channel — A barge and tanker collision caused 252,000 gallons of slurry oil to spill into the Ship Channel.

6/24/89 Delaware River Oil Spill — A Uruguayan tanker spilled 300,000 gallons of fuel oil as it ran aground in the Delaware River.

10/23/89 Phillips Petroleum Pasadena Explosion — Gas release led to explosion which destroyed a portion of a polyethylene plant. There were 23 fatalities and 314 injuries. Phillips experienced a $431 million decrease in net income that year. OSHA first proposed $6,400,000 in fines. Later reduced them in exchange for their (Phillips') promise to institute process safety management procedures at Pasadena and three other plants.

Ongoing: Phillips Petroleum's President Glenn Cox said the company's own investigation showed the explosion " ... was the result of a departure from established routine procedures..."

1/1/90 Arthur Kill Oil Spill — Exxon refinery pipeline spilled 567,000 gallons of heating oil. The under-water pipeline running from an Exxon refinery ruptured and spilled the oil into Arthur Kill.

2/7/90 Southern California Oil Spill — American Trader (charted by British Petroleum) spilled 400,000 gallons of Alaskan crude oil. The ship was "gored" by its own anchor!

6/90 Mega Borg Tanker Spill — A tanker spilled 38,000,000 gallons of light crude into the Gulf of Mexico, near the shoreline of Galveston Island. An explosion had occurred aboard the super-tanker.

7/90 Shinoussa Tanker Spill — A tanker spilled 700,000 gallons of crude oil into Galveston Bay. The tanker and barge collided in the Houston Ship Channel.

7/90 Cincinnati Resins Plant Explosion — The BASF plant suffered an explosion and fire. Workers had a few seconds warning. A flammable solvent venting into the plant may have been the cause. There was one fatality and 56 injuries. $1,100,000 in OSHA fines.

7/5/90 Arco in Channelview, Texas Chemical Plant Explosion — Houston Arco plant suffered an explosion that burned a city block-sized area. The fire lasted over four hours. Inadequate training and excessive overtime work have been mentioned as possible causes of the accident. There were 17 fatalities. $3,480,000 in fines.

1991 Philadelphia Electric Company — Nuclear Plant error was reported by the Nuclear Regulatory Commission. Operators were found sleeping in control centers. ·$1,250,000 NRC fine. Also: $500 to $1,000 fines to 33 workers and supervisors.

1991 Staten Island Ferry Terminal Fire — Unauthorized people (smoking crack?) in the building probably set off the fire in an 8-foot-high cockloft. $50,000,000 was estimated to rebuild the terminal building.

3/3/91 Citgo, Lake Charles Refinery Explosion — The refinery experienced a major explosion and fire. There were six fatalities and six injuries. $5,800,000 in OSHA fines. Citgo promised to enhance safety at Lake Charles and two other locations.

5/1/91 IMC Nitroparaffins Plant Explosion — IMC Fertilizer operated the plant for Angus Chemical. There were eight fatalities, 42 workers injured and 70 residents injured. $9,800,000 IMC Fertilizer Inc. OSHA fine and $200,000 Angus Chemical OSHA fine. Both companies also promised to institute or upgrade process safety management procedures at four IMCF/Angus and IMCF facilities. The analyses must cover human error and startup-shutdown conditions.

4/13/92 Chicago Tunnel Flood — One hundred year-old freight tunnel ruptured and spilled millions of gallons of water into downtown Chicago. Although he saw the damaged wall, an engineer did not react and "waited for photos from the drugstore."

On-going. The Mayor fired at least two officials and suspended others who apparently knew about the damage. Inspectors looked at only one bridge because "parking was difficult."

Other cases to follow include a 1992 New York Emergency Room incident. Possibly a milestone legal case where two doctors treated a patient and made serious mistakes. They had worked 18 hours straight. Regulations may be proposed to limit on-duty hours.

### Incidents and Costs Endnote

[1]This table of sample incidents and costs was generated from accessible CNN, AP, safety, and press reports and is incomplete. We do not claim to maintain the absolutely latest data and would appreciate updates from industry. The costs are estimated by journalists, court reports, and the businesses involved. At some later date, we will develop a more representative sample set where more transportation incidents and country-/world-wide incidents are reported.

# APPENDIX THREE

## Additional Resources

*Fundamentals of Shiftwork Scheduling* by James C. Miller, Ph.D., 1992, 59 pages. Available from Miller Ergonomics, 8915 Rocket Ridge Road, Lakeside, CA 92040 Fax: 619-443-6824.

Night and Shiftwork Long-Term Effects and Their Prevention – Proceedings of the VII International Symposium on Night and Shiftwork, edited by R. Cervinka, Austria, 1985.

*Principles and Practice of Sleep Medicine* by Meir H. Kryger, Thomas Roth, and William C. Dement, W. B. Saunders Company, 1989. (Chapter 32: Jet Lag and Sleep Disruption by R. Curtis Graeber.)

*Rhythmic Aspects of Behavior* by R. Curtis Graeber, edited by Fredrick M. Brown, published by Lawrence Erlbaum Associated, 1982.

*Some Must Watch While Some Must Sleep* by William C. Dement, Stanford Alumni Association, 1972.

*Studies of Shiftwork*, Edited by J. Rutenfranz, Taylor and Francis Publishers, 1980.

*The Human Aspects of Shiftwork* by the Institute of Personnel Management, Brookfield Publishing Company, England, 1978.

*Toward a Self Managed Life Style* by Robert L. Williams and James D. Long, Houghton Mifflin, 1975.

# Glossary

**Achievement Maintenance Phase**

The proactive life and workstyle of the extra-achievement employee who doesn't fall into burnout. The extra-achiever thrives on 24-hour work.

**Apollo Program**

NASA's program of conquering space flight to the moon.

**Body Clock**

Same as the internal (circadian) clock.

**Burnout Phase**

The third phase of the burnout model. The individual feels capped in his job and in life. The Double H Syndrome is in high gear at this time.

**CapCom**

Job for a fellow astronaut who communicates from Houston, Texas' Mission Control with the space team 24-hours a day.

**Circadian Rhythm**

The daily temperature, alertness, eating, and sleep rhythm found in living beings.

**Critical Job**

A technically empowered job where there is the potential to perform great good or great harm. Typical critical jobs are Nuclear, Chemical, and Refinery Operators, Police Officers, Dispatchers, Firefighters, Cockpit Crews, Hospital Personnel, Doctors, Emergency Response Personnel. Most shiftworkers work in critical jobs. These jobs, which have been neglected in America, require excellent training and constant attention to performance improvement.

### Continuous Performance Improvement

The continual improvement of performance through a tracking and improvement program. F.A.S.T. Tracking is the commercial program that is run in plants and departments.

### Danger Zone

One day and one night Danger Zones occur each twenty-four hours. Approximately at 2 to 5 p.m. and 2 to 5 in the morning. The Danger Zones are periods of lower body temperature and alertness which correspond with higher accident rates in many studies. These are also termed the "zones of vulnerability."

### Darkside Team Morale

Envelopes the entire team or department when the darkside has infected a critical mass of employees. Nothing can please the darkside group with this condition! (This can include members of supervision and management!)

### Darkside View

The darkside view of the shiftworker is held by many managers and shiftworkers themselves. Shiftworkers with this view will have more problems in the areas of fitness, alertness, sleep, time off and general attitude during their shiftwork career. The darkside view is a self-fulfilling prophecy that leads to burnout and the Double H Pattern.

### Divorce Epidemic

Occurs among 24-hour employees in large departments where sufficient attention is not paid to the F.A.S.T. (fitness, alertness, sleep and time off) areas.

### Double H Pattern

The Double Hs of hopeless and helplessness appear ironic in shiftworkers who once began with such great drive and hopes of extra-achievement! This phenomenon, appearing between the third and tenth year is documented by various researchers and confirmed by many shiftworkers and managers.

**Edison, Thomas Alva**

An extra-achiever and inventor of the light bulb who dreamed of a 24-hour, highly productive world.

**F.A.S.T. Tracking**

Our hard-charging performance tracking and improvement program that focuses on fitness, alertness, sleep, and time-off efforts. It is kicked-off by a TIME TRAVELERS program carrying the NASA astronaut theme.

**Fatigue**

Tiredness. Slowing down of the employee due to exertion or sleep deprivation — or both. Also, depressed individuals mention experiencing high fatigue.

**Fumbling Slips**

A class of human error. Those are physical "motor errors" that are easy to observe: mis-sorting, mis-labeling, mis-operating — usually caused by mistakenly touching/ grabbing/ flipping/ adjusting the wrong switch, lever, button, or other object. Fumbling slips are often associated with high pressure (stress).

**GlobeZilla**

The monster of global competition and stress. GlobeZilla is pounding the world of the nineties even more heavily than Glenn had predicted in 1988. GlobeZilla promotes human error and its Set-Ups.

**Hire-In Phase**

The initial phase of both the burnout and achievement models. The employee tries for extra-achievements by hiring into shiftwork.

**Human Error**

Human error is error performed by the individual or group. As researchers and trainers, we divide the error into categories: "SlipWalking error" (individual Attention and Autopilot Slips) and team/communications error. This last type of error includes planning/engineering and intentional error. Most errors are preceded by Set-Ups.

**Human Factors**

Human Factors are those contributions of the human to the incident as well as outside influences on the human performer. Communication and group interaction are included here. The study of human factors is a new area for most industries.

**Human Factors Engineering**

Human Factors Engineering or Ergonomics is the application of human factors principles to humans and their environment, both physical and behavioral. When workers and their environments are mismatched, injury levels increase, production decreases and other incidents occur which lower organizational efficiency and worker welfare.

**Individual Responsibility in Safety**

Individual responsibility in safety has become a strong safety concept recently. The focus can be on learning about surrounding hazards, taking an active role, and intervening during unsafe acts.

**ICE Links**

One unit of human error in a chain of human error. (ICE is the abbreviation for "incredible chain of events.") Often set up by an earlier error and setting up a future error! Intervention in one ICE Link will prevent the final incident from occurring. Useful in incident analysis and incident prevention training. Slips fit into one class of ICE Link.

**Internal Clock**

An "internal clock" has been discovered by sleep researchers to control our daily temperature, sleepiness, and alertness cycle (the circadian cycle).

**Irritability**

A common complaint of shiftworkers, especially those on rotating shifts and experiencing the Double H Pattern.

**Jet lag**

A condition where the internal human clock gets out of synch with the actual time zone. This can be caused by rapid jet travel or by rapidly rotating the hours of sleep/awakeness.

**Lifestyle**

The style in which a person lives. This is sometimes proactively planned. With others, it is a simple reaction to whatever life brings.

**Minimum Compliance Syndrome**

Training corporate personnel at the minimum level that will meet government standards. We use it in a somewhat derogatory manner for those companies that are not getting on top of safety problems.

**Near Miss**

A near-disaster! Most often caused by a human slip. It is usually prevented by an intervention.

**NMAC**

A near midair collision (that is reported in aviation studies).

**Performance**

A combination of one's efforts and the results of those efforts. In business and technology, the most efficient performance is often graded as the best performance, i.e., the least amount of effort resulting in the greatest result. F.A.S.T. Tracking partially concurs with this. However, we believe in maximizing efforts to take performance past previous "bell curves" and boundaries.

**Pimping**

Performing an intervention purposefully in front of the boss. (What next?)

**Primary Support System**

The family and friends of the 24-hour employee.

**Reaction Phase**

The middle phase of the burnout model. Attitude change and reactions to stress are noticeable. The employee begins to feel "in a rut." This phase is fueled by fatigue and jet lag. It is usually caused by endless rotations around the clock.

**Safe Working Style**

A major program researched and written for the Dow Chemical Company. Spin-offs of the program include new insight into human error, incident investigation, and safety intervention.

**Set-Ups**

Precursors (causes) of Human Error Slips.

**Shiftwork Set-Up**

Set-ups that lead to Shiftwork Slips: Poor fitness and health; inappropriate meals; insufficient exercise; stimulant abuse; poor sleeping environments; poor sleep schedules; poor work schedules and excessive overtime; monotony; and distressful family life and off-time.

**Sleep Deficit**

Insufficient sleep building into a deficit. Leads to micro- and macro-sleep. Napping with a sleep deficit may prove dangerous due to sleep inertia.

**Slips**

Single instances of human error in the mental and motor area. We divide slips into various categories, including: Shiftwork Slips (those happening during shiftwork), Attention (Switching and Forgetting) and AutoPilot (Fumbling and Robotic) Slips. SlipWalking is the term we use for all of these mental and motor slips. Slips can usually be traced to a Set-Up.

**SlipWalking**

SlipWalking is committing inadvertent Slips due to autopilot (mindless habituation) or inattention. These Slips include mis-labeling, mis-sorting, and mis-operating. Both ends of the boredom-emergency spectrum can cause SlipWalking.

**Spin-Off Result**

A second or unexpected result of performing an effort.

**Technical Empowerment**

Empowerment through technology. For example: size, complexity, computerization, horsepower, speed, toxicity, high pressure, and flammability.

**Views of the Shiftworker**

Two extreme views of the shiftworker are prevalent in America: The Darkside View and the Brightside view. These become self-fulfilling prophecies — leading to the two shiftwork models: Burnout and Achievement.

# Index

## —U—



## —V—



## —W—